Haptic Interaction with Deformable Objects

For further volumes:
www.springer.com/series/8786

Springer Series on Touch and Haptic Systems

Series Editors

Manuel Ferre
Marc O. Ernst
Alan Wing

Series Editorial Board

Carlo A. Avizzano
José M. Azorín
Soledad Ballesteros
Massimo Bergamasco
Antonio Bicchi
Martin Buss
Jan van Erp
Matthias Harders
William S. Harwin
Vincent Hayward
Juan M. Ibarra
Astrid Kappers
Abderrahmane Kheddar
Miguel A. Otaduy
Angelika Peer
Jerome Perret
Jean-Louis Thonnard

Guido Böttcher

Haptic Interaction with Deformable Objects

Modelling VR Systems for Textiles

 Springer

Dr. rer.nat. Guido Böttcher
Institut für Mensch-Maschine Kommunikation
Fachgebiet Graphische Datenverarbeitung
Gottfried Wilhelm Leibniz Universität
Hannover
Welfengarten 1
30167 Hannover
Germany
boettcher@welfenlab.de

ISSN 2192-2977   e-ISSN 2192-2985
Springer Series on Touch and Haptic Systems
ISBN 978-1-4471-2684-3   ISBN 978-0-85729-935-2 (eBook)
DOI 10.1007/978-0-85729-935-2
Springer London Dordrecht Heidelberg New York

British Library Cataloguing in Publication Data
A catalogue record for this book is available from the British Library

Cover design: VTeX UAB, Lithuania

Printed on acid-free paper

Springer is part of Springer Science+Business Media (www.springer.com)

Series Editors' Foreword

This is the third volume of the "Springer Series on Touch and Haptic Systems", which is published as a collaboration between **Springer** and the **EuroHaptics Society**. This book is focused on haptic interaction with deformable and textile objects. This area represents one of the most currently advanced topics in haptic interaction, given the confluence of a range of issues including, synchronisation of visual and haptic excitation, high-speed solution of complex equations and development of dynamic scenarios. All these matters have been analysed in depth by the author. He provides an excellent panorama of the current state of the art in the use of haptic interaction for deformable objects.

The physical simulation has been explained in detail by the author. The problems about modelling deformable object are shown in chapter two, which require complex mathematical formulations in order to obtain realistic haptic interactions. In chapter three, methods for haptic rendering and collision detection are broadly described, showing examples of applications. The development and analysis of a complete virtual reality scenario are explained in chapters four and five. This work provides readers valuable information about the main components and performance required to interact with deformable objects using haptics and graphical simulations.

Manuel Ferre
Marc O. Ernst
Alan Wing

Preface

In the beginning of 2005, right after successful completion of my mathematical studies my professor asked me if I would like to continue my research in the haptic rendering by participating in an EU-project. As the project named HAPTEX involved several academic partners collaborating to reach a final very ambitious goal of feeling virtual textiles, I was immediately attracted by the idea working together with other international researchers. Moreover, as I already wrote my diploma thesis to some extent in the field of haptics, I was eager to extend my knowledge to create more sophisticated rendering algorithms. Especially the combination of physical simulation with real-time haptics motivated me as I am very interested in physics and graphic programming.

Since the start of the project in the same year I met experts of various fields, who shared their knowledge with me. Studying mathematics, I was rather focused on theoretical ideas with limited practical applications. In this project I gained some broader view of the mathematical tools and their use in different fields including mechanical engineering as well as control theory. With the text at hand I want to do the same as my partners did in sharing my gained knowledge with the reader. Its content is therefore directed to students with background in mathematics and computer science like me in the beginning of the project.

The monograph would have never been possible without the funding of the aforementioned project "HAPtic sensing of virtual TEXtile" (HAPTEX) under the Sixth Framework Programme (FP6) of the European Union (Contract No. IST-6549). The funding was provided by the Future and Emerging Technologies (FET) Programme, which is part of the Information Society Technologies (IST) programme and focuses on novel and emerging scientific ideas. For bringing the project to life (and thus requesting the funding) I thank Prof. Nadia Magnenat-Thalmann, Dr. Harriet Meinander, Prof. Franz-Erich Wolter, Dr. Ian Summers, and P.Eng. Fabio Salsedo.

At the same time I want to thank my doctoral adviser Prof. Franz-Erich Wolter for having this great opportunity to work in such an interesting field and to participate in the project. Moreover, I would also like to thank Prof. Nadia Magnenat-Thalmann for her effort to coordinate the project. Without her pushing all members to the limit, we would have never been successful in reaching the goal. Certainly, a project lives from good collaboration and this was clearly given by the partners. I especially appreciated the discussions with Dr. Pascal Volino. I thank him for gaining insights in the modelling of textile mechanics and its efficient simulation.

I was also happy to collaborate with my long time working colleague Dennis Allerkamp in the project. He concentrated his research on the tactile rendering of the textile surfaces. He gave me not only moral support in our stressful work but also directed our interests towards the important things. I owe him a lot by his drive at work.

I will not forget to mention my students who were extremely helpful by their support in the development of the system. With the time pressure of the intermediate project goals I would not have succeeded without them. I therefore thank Rasmus Buchmann, Michael Hanel, Maximilian Müller and especially Steffen Blume and Daniel Glöckner who worked very hard on the software development.

Finally, I would have never finished my work without the support of my family and my wife. I also highly appreciated the work of my colleagues and friends to proof-read my manuscript. Thank you very much for your endurance.

Hannover, Germany   Guido Böttcher

Contents

Acronyms

AABB	axis-aligned bounding box
BS	bounding sphere
BVH	bounding volume hierarchy
CG	conjugate gradient
CPU	central processing unit
DES	differential equation system
DOP	discrete oriented polytope
DFT	discrete Fourier transform
FIR	finite impulse response
FFT	fast Fourier transform
FEM	finite element method
GRAB	graphics access for blind people
IIR	infinite impulse response
ISA	industry standard architecture
KES-F	Kawabata evaluation system for fabrics
LES	linear equation system
LCP	linear complementarity problem
MBS	multibody System
PCG	preconditioned conjugate gradient
TOH	total harmonic distortion
VR	virtual reality
ZOH	zero order hold

Chapter 1
Introduction

In the past decades computer graphics has become an essential part in science as it provides algorithms and methods helping to visualise problems and processes in different fields of research. For example, in weather forecasts, models are used to simulate the dynamics of the atmosphere and its interaction with seas and land masses. Without an adequate visualisation of the huge data sets produced by the models, it is very hard to make a forecast relying on the computed data.

By giving the user a visual representation of the computer-generated output it is possible to instantly understand what has been computed and how it is related to a stated problem. This approach has been driven further to a level where the user is completely surrounded by a virtual environment which resembles reality. Special stereoscopic displays create a true three-dimensional view that is perfecting the illusion to be fully immersed in this so-called Virtual Reality (VR).

While the graphical rendering of virtual environments is more and more indistinguishable from real images, the interaction inside of such a world is quite far away from being realistic. Several aspects in the interaction between the user and VR are still missing. The most important aspect is the haptic feeling when a user is in touch with an object. Even though visual sense is mainly relied on, the sensation of touching an object is still needed. Otherwise, intuitively grasping and manipulating objects inside VR could not be done. Moreover, force interaction occurring at the contact has to be felt. Generally speaking, when an object is grasped, an energy transfer from the user to the object and vice versa is obtained. It means that the effect of the energy transfer in the mechanical behaviour of the object touched with respect to the underlying material has to be considered. In case of a rigid object causing a change of potential and kinetic energy, the transfer is visible in the motion, whereas for a soft body, it results in a deformation under the load the user applies at the contact. A VR system capable of dealing with such physical processes has to make sure that it delivers the contact forces at a very high update rate (approximately 1 kHz). This high update rate is necessary to avoid stability problems in the control of force-feedback devices and in order to achieve a high fidelity haptic sensation in the interaction.

Although in previous works many issues were solved in previous works especially in contact descriptions, touch interaction many issues were solved in previous

G. Böttcher, *Haptic Interaction with Deformable Objects*,
Springer Series on Touch and Haptic Systems,
DOI 10.1007/978-0-85729-935-2_1, © Springer-Verlag London Limited 2011

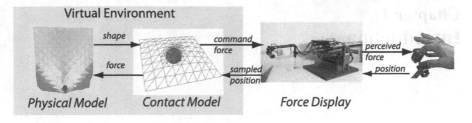

Fig. 1.1 Signal loop in haptically augmented VR system

works especially in contact descriptions, touch interaction was limited to either low update rates or linear models. Interaction with very light and thin deformable objects was not in the focus of prior research projects. Our approach described in [4] provides a general solution for the haptic simulation of deformable objects by implementing a multi-rate approach with an improved local buffer model (see Fig. 1.1). When two dynamical systems are coupled, usually the issue of synchronisation and concurrency usually arises. In our case, one system has the constraint of high response demanded by haptics, thus the time spent in computing the system state has to be controlled carefully. Additionally, considering available computation power, the synchronisation should minimise the dead time originating from synchronisation (cf. Böttcher et al. [5]).

1.1 Previous Work

Integrating the sense of touch in a VR system is not simply an addition to visual rendering. Since touch perception is very different from other perceptional channels, requirements in hardware and software are much higher to provide an acceptable illusion. Our sense of touch provides a unique and bidirectional link to the environment. The requirements in having a realistic impression of touching virtual objects belong to the two main issues still present in today's VR systems: the stability of touch feedback and the need for short response times of approximately 1 ms. The latter was ensured by a fast and simple algorithm proposed by Zilles and Salisbury [19]. The algorithm generates a touch feedback by computing a spring force proportional to the penetration depth of the user in the touched object. Ruspini et al. [16] presented an improved algorithm which solved problems in some geometric configurations of the prior algorithm and displayed surface features. Both algorithms were conceived to provide interaction forces with rigid bodies.

At the same time, Bro-Nielsen and Cotin [6] showed a visually interactive simulation of linear finite element (FE) models for medical applications. Subsequent to the latter contribution, Astley and Hayward [1] proposed an approach to apply these results to touch feedback. In order to distribute the computational load and to overcome the issue of much higher response time, a decoupling of the touch computation and model simulation by using a multi-layered FE mesh evaluated at different update rates (multi-rate) was proposed. However, this implementation yielding 10 Hz was far below the desired response times of touch perception. Although the model was not applicable to touch feedback, the concept of multi-rate models became quite popular. A similar approach was made by Balaniuk [2] who approximated the actual geometry at the contact of a deformed body by a sphere. This static spherical region served as a "local buffer" for the touch feedback until the simulation updated geometry changes.

As the new multi-rate models, respectively intermediate models, feed forces from touch interaction back to the global model running at a lower rate, researchers began to analyse their behaviour in terms of stability. Cavusoglu and Tendick [7] analysed the exchange at multi-rate in a simplified situation by a single non-linear spring with different couplings. A similar analysis was made by Barbagli et al. [3] but with multi-contact. They found out that stability is preserved when the model stiffness is limited w.r.t. the multi-rate approach. Lee and Lee [12] proposed a non-linear virtual coupling between the models to achieve stability. Kim et al. [11] proposed an algorithm ensuring passivity, respectively stability, of the multi-rate system by bounding the energy induced by delays.

A different way without intermediate models was proposed by Zhuang and Canny [18]. A single non-linear FE model displayed with time delay in graphics and touch was simulated. Considering such point, constraining the simulation to react on touch contact was able to be done. Short touch response was achieved by force interpolation. Mazzella et al. [15] proposed an improvement to the interpolation by buffering previously computed interaction forces which are linearly composed w.r.t. the distance of the contact point. A different approach from Mahvash and Hayward [14] uses a precomputed force response of the model approximated by polynomial splines to create a touch feedback at high update rates. James and Pai [10] had a similar approach with an enhanced contact description by a traction field, but it was limited to linear elastostatic models.

The most sophisticated and most flexible approach in touch feedback has been proposed by Duriez et al. [8]. The method is based on the Signorini's contact problem commonly used in contact mechanics to model objects in contact. As a result of the comprehensive treatment of the contact it requires much computation time and cannot satisfy haptic real time requirements by reaching only 100 Hz. The authors close the time gap by using the force Jacobian to create an estimate of the force evolution in relation to the positional change.

Although with the previous work many issues especially in contact description were solved, the touch interaction was limited to either low update rates or linear models. More importantly, interaction with very light and thin deformable objects was not in the focus of prior research projects.

The work tries to close this gap in resembling the assessment of textiles. With a multi-rate approach for haptic rendering together with an improved local buffer model, a non-linear simulation of the textile contact is achieved in real time. A special two-finger contact model is conceived to provide the touch feedback for the aforementioned type of objects (cf. Böttcher et al. [4]).

With the EU-funded HAPTEX project, coordinated by the MIRALab at the University of Geneva, an effort was made in providing methods and models for visual

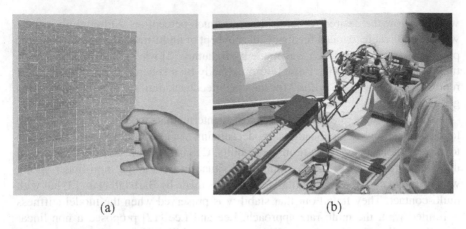

Fig. 1.2 (a) Interaction Scenario: Touching and stretching textiles. (b) Final demonstrator and interaction scenario of the HAPTEX System

and haptic realism which integrated tactile and force feedback for the first time (cf. [9, 13, 17]). The presented VR system consequently used some soft- and hardware components of the project partners. The MIRALab contributed a real time simulation model for textiles. The SmartWearLab at the University of Tampere provided a selection of fabrics to be simulated by the system which featured important palpable characteristics of fabrics. For displaying the very small forces involved in the textile touch, special force-feedback hardware was developed by the PERCRO Laboratory at the Scuola Superiore di Sant'Anna. The Biomedical Physics Group at University of Exeter developed actuators called *Tactile Arrays* creating tactile stimulations allowing tactile feedback of the textile surface.

This work provides the knowledge to create a VR system that is able to reproduce multimodal perception of touching virtual textiles. The application scenario of the VR system is the kinesthetic and tactile exploration of a virtual piece of fabric hanging on a rigid support. To provide the touch perception it uses the special force feedback hardware with an integrated tactile array stimulating the fingertips (illustrated by Fig. 1.2(b).

Since the forces generated by the software are based on the physical simulation, the contact model, and the force-feedback device, it is necessary to consider the effects of each component influencing the perception. Consequently, the work at hand reviews the numerical modelling of continuous materials and analyses a force feedback device with its control problems.

It starts in Chap. 2 with a thorough introduction to simulation of mechanical systems. Beginning with the fundamental laws of Newton's kinematics and subsequently the definition of multibody systems a proper derivation of the physical models governing continuous media is given. Special attention is further given to material behaviour in general and in particular of fabrics. Together with the complete physical description, numerical methods are presented solving the respective equations of the physical model to simulate its dynamics.

Since the VR system can be seen as a control loop with the renderer and the haptic device as signal processors, the VR system's stability depends on its signal responses. In Chap. 3, the principles of control and response of haptic devices are described. Together with the tools for contact detection, the prerequisites for haptic rendering are at hand.

In Chap. 4 the VR system software framework developed in this thesis is presented. In the description and explanation of each component algorithms are presented ensuring the necessary real time latency. Furthermore, a contact model is described that allows the grasp of a textile with two fingers. Additionally, the multirate approach for separating the computations of contact area from the complete textile simulation for the haptic rendering is illustrated.

Chapter 5 analyses the VR system in terms of haptic quality. By measuring the frequency response and the accuracy of the used haptic hardware, the corresponding limitations of realistic haptic rendering in the given context are identified. The rendering itself is to some extent verified by the definition of standard user interaction showing the mechanical characteristics of the materials in the rendering. With a subjective evaluation of different virtual textiles an assessment of the VR system is made.

Finally, Chap. 6 gives a summary of the work and its results. Moreover, suggestions for future investigations and the remaining problems are given.

References

1. Astley, O., Hayward, V.: Real-time finite elements simulation of general visco-elastic materials for haptic presentation. In: IEEE/RSJ International Conference on Intelligent Robots and Systems (IROS). IEEE Computer Society, Los Alamitos (1997)
2. Balaniuk, R.: Using fast local modeling to buffer haptic data. In: PUG99: Proceedings of the Fourth PHANTOM Users Group Workshop (1999)
3. Barbagli, F., Salisbury, K., Prattichizzo, D.: Dynamic local models for stable multi-contact haptic interaction with deformable objects. In: 11th Symposium on Haptic Interfaces for Virtual Environment and Teleoperator Systems, 2003. HAPTICS 2003. Proceedings, pp. 109–116 (2003). doi:10.1109/HAPTIC.2003.1191248
4. Böttcher, G., Allerkamp, D., Glöckner, D., Wolter, F.E.: Haptic two-finger contact with textiles. Vis. Comput. 24(10), 911–922 (2008)
5. Böttcher, G., Allerkamp, D., Wolter, F.E.: Multi-rate coupling of physical simulations for haptic interaction with deformable objects. Vis. Comput. 26(6), 903–914 (2010)
6. Bro-Nielsen, M., Cotin, S.: Real-time volumetric deformable models for surgery simulation using finite elements and condensation. Comput. Graph. Forum 15(3), 57–66 (1996)
7. Cavusoglu, M.C., Tendick, F.: Multirate simulation for high fidelity haptic interaction with deformable. In: IEEE International Conference on Robotics and Automation, 2000. Proceedings. ICRA'00, vol. 3, pp. 2458–2464 (2000). doi:10.1109/ROBOT.2000.846397
8. Duriez, C., Andriot, C., Kheddar, A.: Signorini's contact model for deformable objects in haptic simulations. In: 2004 IEEE/RSJ International Conference on Intelligent Robots and Systems, 2004 (IROS 2004). Proceedings, vol. 4, pp. 3232–3237 (2004). doi:10.1109/IROS.2004.1389915
9. Fontana, M., Marcheschi, S., Tarri, F., Salsedo, F., Bergamasco, M., Allerkamp, D., Böttcher, G., Wolter, F.-E., Brady, A.C., Qu, J., Summers, I.R.: Integrating force and tactile rendering into a single vr system. International Conference on Cyberworlds, 2007 CW '07 (24–26 Oct. 2007), pp. 277–284 (2007). doi:10.1109/CW.2007.40

10. James, D.L., Pai, D.K.: A unified treatment of elastostatic contact simulation for real time haptics. In: ACM SIGGRAPH 2005 Courses, p. 141. ACM, New York (2005)
11. Kim, J.P., Seo, C., Ryu, J.: A multirate energy bounding algorithm for high fidelity stable haptic interaction control. In: SICE-ICASE, 2006. International Joint Conference, pp. 215–220 (2006)
12. Lee, M.H., Lee, D.Y.: Stability of haptic interface using nonlinear virtual coupling. In: IEEE International Conference on Systems, Man and Cybernetics, 2003, vol. 4 (2003)
13. Magnenat-Thalmann, N., Volino, P., Bonanni, U., Summers, I., Bergamasco, M., Salsedo, F., Wolter, F.: From physics-based simulation to the touching of textiles: the haptex project. Int. J. Virtual Real. 6(3), 35–44 (2007)
14. Mahvash, M., Hayward, V.: High-fidelity haptic synthesis of contact with deformable bodies. IEEE Comput. Graph. Appl. 24(2), 48–55 (2004). doi:10.1109/MCG.2004.1274061
15. Mazzella, F., Montgomery, K., Latombe, J.C.: The forcegrid: a buffer structure for haptic interaction with virtual elastic objects. In: IEEE International Conference on Robotics and Automation, 2002. Proceedings. ICRA'02, vol. 1, pp. 939–946 (2002). doi:10.1109/ROBOT.2002.1013477
16. Ruspini, D.C., Kolarov, K., Khatib, O.: Haptic interaction in virtual environments. In: Proceedings of the 1997 IEEE/RSJ International Conference on Intelligent Robots and Systems, 1997. IROS '97, vol. 1, pp. 128–133 (1997). doi:10.1109/IROS.1997.649024
17. Salsedo, F., Fontana, M., Tarri, F., Ruffaldi, E., Bergamasco, M., Magnenat-Thalmann, N., Volino, P., Bonanni, U., Brady, A., Summers, I., Qu, J., Allerkamp, D., Böttcher, G., Wolter, F.-E., Mäkinen, M., Meinander, H.: Architectural design of the haptex system. In: Proceedings of the HAPTEX'05 Workshop on Haptic and Perception of Deformable Objects, Hanover, pp. 17–29 (2005)
18. Zhuang, Y., Canny, J.: Haptic interaction with global deformations. In: IEEE International Conference on Robotics and Automation, 2000. Proceedings. ICRA'00, vol. 3, pp. 2428–2433 (2000). doi:10.1109/ROBOT.2000.846391
19. Zilles, C.B., Salisbury, J.K.: A constraint-based god-object method for haptic display. In: IEEE/RSJ International Conference on Intelligent Robots and Systems 95. 'Human Robot Interaction and Cooperative Robots', Proceedings, vol. 3, pp. 146–151. IEEE Computer Society, Los Alamitos (1995). doi:10.1109/IROS.1995.525876

Chapter 2
Physical Simulation

Without the understanding of the laws behind the physical behaviour of objects by means of interaction, a realistic simulation within a virtual environment cannot be achieved. The fundamental principles of the subject rely on the theories of classical and continuum mechanics. Therefore, in the following a review the physics governing the motion and deformation of bodies is given. The notations of the physical properties are based on [5, 13] and [14] which also give a good introduction, but have broader perspective in the concepts of classical mechanics.

Since the book at hand is not meant to give a complete introduction, it is advised to complement the knowledge by additional literature for a full understanding. The book has to focus on the principles required for the simulation of deformable objects respectively textiles.

For detailed introduction to the concepts of linear respectively for nonlinear elasticity which will be a part of this chapter, the books of Gould [6] and Bonet [3] besides the ones mentioned above can be recommended.

2.1 Elementary Units and Principles

The motion of rigid bodies was one of the first physical phenomena that was scientifically investigated and marks to some extent the beginning of physics. Before this point, all observations and experiments were more dedicated to answer questions on a philosophical level. It started with Galileo Galilee in the late 16th century, who mainly studied the motion of objects in free fall. He experimentally proved that objects of different mass will fall within the same time as long as the resistance of the surrounding medium is negligible. Moreover, he found that a body on a level surface will continue its movement in the same direction at constant speed unless it is disturbed. The latter finding, the principle of inertia, was a great progress in the understanding of motion. Galileo also derived a relationship between the square of the elapsed time to the distance. Although Galileo made many contributions to science, the actual founder of classical mechanics which is the basis of the work is Sir Isaac Newton. He established by his findings and by those of his predecessors

G. Böttcher, *Haptic Interaction with Deformable Objects*,
Springer Series on Touch and Haptic Systems,
DOI 10.1007/978-0-85729-935-2_2, © Springer-Verlag London Limited 2011

a mathematical basis and presented the groundwork for classical mechanics by his publication of the *Philosophae Naturalis Principia Mathematica*.

2.1.1 Conventions

In the following the motion of a body in time will be deduced analytically. First of all, to describe the movement the **position** of the body is needed, denoted by **r** being a vectorial parameter (indicated by boldface letters) and expressed in a general coordinate system with its time parameter t. For now, it is assumed that the body has a negligible size. By observing the positional change of the body in an infinitesimal time step one arrives at the differential of $\mathbf{r}(t)$ given by

$$\dot{\mathbf{r}} = \frac{d\mathbf{r}}{dt}$$

which leads to the **current velocity** measured in $[\frac{m}{s}]$.

Further, it can be seen that the introduction of the body mass as a quantitative measure of the inertia is an immediate result of Galileo's principle. For example: given an object with a certain velocity, a change in this physical value can only be achieved if a force is applied. This force is required to overcome the inertia and thus leads to the following relationship between force in [N] and mass in [kg] (inertia):

$$\mathbf{F} = \frac{d}{dt}(m\dot{\mathbf{r}}) = \frac{d\mathbf{p}}{dt} \equiv \dot{\mathbf{p}} \tag{2.1}$$

In words, a change of the (vectorial) velocity in time requires a force acting on the body. The latter equation constitutes Newton's second law of dynamics. At the same time, another quantity is introduced, the (linear) **momentum p**. It is mathematically expressed by the product of mass and velocity. The common notation $\dot{\mathbf{p}}$ denotes the differential change in time of **p** which is equivalent to the **impulse** $[\frac{kg \cdot m}{s}]$ being the integral of a force w.r.t. time. The momentum can be seen as mass in motion, e.g. to stop a body with a high momentum, a high force will be needed.

In classical mechanics, constant mass over time is normally assumed, simplifying the previous formula to

$$\mathbf{F} = m\frac{d\dot{\mathbf{r}}}{dt} = m\ddot{\mathbf{r}} \tag{2.2}$$

This results in the rate of change in velocity over time, the **acceleration** $[\frac{m}{s^2}]$. The basic relationship between force and acceleration given by the latter equation is considered to be the fundamental principle of Newtonian dynamics. The principle will later on be used as a basis for the physical simulations. A second-order differential equation $\mathbf{F} = \mathbf{m} \cdot \frac{d^2}{dt^2}\mathbf{r}$ is created which has to be solved numerically for the acceleration. Equation (2.1) additionally provides an important concept namely the **principle of conservation of linear momentum**. The law says that at vanishing net force, i.e. $\mathbf{F} = 0$, the momentum is conserved.

With the linear momentum, a quantity is given describing the motion of an object. Furthermore, an object can also feature a rotational motion. This rotation is described by the **angular momentum L** relative to an origin **O**. It is defined by

$$L = (r - O) \times p \qquad (2.3)$$

The rate of change in time of the angular momentum leads us to

$$\tau = \frac{dL}{dt} = \underbrace{\frac{d(r - O)}{dt} \times p}_{\dot{r} \times m\dot{r} = 0} + (r - O) \times \frac{dp}{dt} = (r - O) \times F \qquad (2.4)$$

which is called **torque**. Similarly to the linear momentum, (2.4) yields the principle of conservation of the angular momentum, which states that the angular momentum is conserved if the torque $\tau = 0$.

2.1.2 Energy and Conservation

With the brief introduction of important vectorial units and their relationships, the current mechanical state of a system can be given in terms of forces and velocities. However, these quantities are only the result of a more basic quantity, namely the energy being inherent in a mechanical system. Energy is an abstract term for a scalar quantity to be associated with the system. Nevertheless energy can be categorised in different forms which will be illustrated in the following. In the field of mechanical and structural engineering, the **potential energy** is the most pronounced nature of energy. Typically in balanced mechanical systems, the biggest quantity of the energy is generally conserved in the potential form, e.g. due to its own weight, as there is no movement. The weight corresponds with the gravitational quantity of the potential energy V. It is defined by the mass m of the body and its height h relative to the ground, respectively to the origin along the line of action of the gravity within the inertial system. Therefore it can be described by

$$V = m \cdot g \cdot h \qquad (2.5)$$

It should be noted that the formula is just an approximation, as the introduction of the gravitational acceleration g is only sufficiently accurate at ground level. The previously shown relationship is also a special case of potential energy, i.e. in a more general form (independent of its physical nature) it can be written as a scalar field

$$V : \begin{cases} \mathbb{R}^3 \to \mathbb{R} \\ r \mapsto V(r) \end{cases} \qquad (2.6)$$

which maps the object coordinates to an energy level. A fundamental aspect is the **preservation of energy** governing all known natural phenomena. It states that *the*

total energy of a physical (closed) system is always conserved and thus constant
through all state transitions of the system.

The term *closed* in this context has the meaning of complete isolation from outside such that energy can neither drain nor flow in. A direct result of this basic principle is the conservation of the momentums already introduced above in (2.1) and (2.4). The conservation of energy is not only a physical requirement but also a tool allowing to define the solution of a physical problem mathematically which will be made use of later.

Another term in this topic is the **work** produced inside a system. It is in contrast to the "common" energy-term used to define the quantities needed or produced in mechanical processes where a force is acting. Since in a system being in an equilibrium such a process cannot start without external influence, a form of mechanical energy has to be provided over a time or a path which is then normally expressed in the terms of work. Work is therefore achieved by applying an external force **F** during a displacement or along a path S. Therefore the formal definition of work measured in (N m) W_{ab} along the path is

$$W = \int_S \mathbf{F} \cdot d\mathbf{s} \tag{2.7}$$

The work in only one dimension is observed first, generated in the movement from point a to b. If one looks at an object and inserts the resulting force (2.2) into the above equation, the following formula for the work can be obtained.

$$W_{ab} = \int_a^b m\ddot{\mathbf{r}} \cdot d\mathbf{s} = m \int_a^b \frac{d\dot{\mathbf{r}}}{dt} \cdot d\mathbf{s} = m \int_a^b \frac{d\dot{\mathbf{r}}}{dt} \cdot \dot{\mathbf{r}}\, dt = \int_a^b \dot{\mathbf{r}}\, d\dot{\mathbf{r}} \tag{2.8}$$

$$= \frac{m}{2} \int_a^b \frac{d}{dt}(\dot{r}^2)\, dt = \frac{m}{2}(\dot{r}_b^2 - \dot{r}_a^2) \tag{2.9}$$

From this result it can be seen that the work solely depends on the scalar quantity $m\dot{r}^2$ which is computed at the points a and b. This specific quantity being the product of mass and its squared velocity is called the **kinetic energy** of a mass point. It is different from the potential energy by being preserved in the motion of the body. It is therefore also known as the motion energy due to the velocity. With the previous observation expressed in formula (2.9) it is said that the work done can be defined as the difference of kinetic energy at the two points. It simply be written as

$$W_{ab} = T_b - T_a \tag{2.10}$$

Although the path is not explicitly given in the formula, the work done still depends on it. Thus another consideration has to be taken into account: if all arbitrary paths as illustrated in Fig. 2.1 taken from a to b lead to the same amount of work done, the force and the system are said to be **conservative**.

One can arrive at an equivalent conclusion by describing a conservative force by using a closed path. If the work done on an arbitrary path from a to b and back to a vanishes, then the force satisfies the conservation condition. This will always

Fig. 2.1 Examples of
physical paths connecting two
points

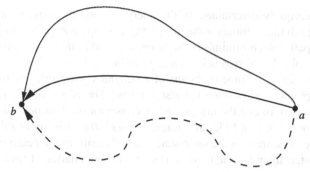

be true for a homogeneous gravitational field. But adding a dissipative component like friction to the same field would break this condition, as the force gets a positive component that does not vanish. Formally, with aforementioned property

$$\oint_a^b \mathbf{F}\,d\mathbf{s} = 0 \qquad (2.11)$$

holds. If one uses the conclusions of the "fundamental theorem of calculus" and the "gradient theorem" force can be expressed by a scalar field. An essential and sufficient condition for the existence of such a scalar field is the independence of the path. Then, the scalar function w.r.t. position fulfils the condition of the potential field of (2.6), and one obtains the following relation

$$\mathbf{F} = -\nabla V(\mathbf{r}) \qquad (2.12)$$

One should note that the negative sign is physically motivated by V being an energy. To be more precise: In vector calculus the gradient is defined as the steepest increase of a function and to obtain the physical property of energy conservation, the force must always be opposing the direction of increase of energy. Thus, it follows for the work

$$W_{ab} = \int_a^b -\frac{\partial V}{\partial s}\,ds = V_a - V_b \qquad (2.13)$$

As one can see from the equation above, the work is independent of the choice of field origin, as it does not change the value. Combining the results of (2.10) and (2.13), one gets

$$T_b - T_a = V_a - V_b$$

or equivalently

$$T_b + V_b = V_a + T_a \qquad (2.14)$$

The result can be seen as a mathematical proof for the energy conservation since the net energy, given by $E = T + V$, is retained in a conservative potential field.

  As a consequence discovered by Euler and Lagrange, given a particle with known velocity and position at start and further the end position, then the actual path is uniquely determined by the energy E and its conservation. This allows to find a path that minimises the time-integral of the energy E along the path. The particular path which minimises the integral called *action* is then the path chosen by nature. This observation is known as *principle of least action*.

But for some problems of mechanics, the work function (2.13) is not only dependent on position but also on time. Therefore the scalar field is not conservative anymore and the law of energy conservation does not hold. Hamilton extended the procedure of Euler and Lagrange such that it is applicable for the latter case. The procedure uses the same start and end positions but determines further the time from start to end of the motion. Hamilton's formulation of least action is given by

$$\delta \int_{t_1}^{t_2} (T - V)\, dt = 0 \qquad (2.15)$$

which assert that the action assumes its minimum for motion taken by the particle. The principle of least action paved the way for a variational approach to the motion of an object which will be further discussed in the continuum mechanics in Sect. 2.2.

2.1.3 Multibody Systems and Constrained Motions

Having formulated a framework for deducing the laws governing the motion of a single body and introduced the important physical quantities, it is now time to generalise the mechanics to multibody systems (MBS). As the name implies, these systems are consisting of many bodies or mass points, that are acted upon by a common force. In the following, it is first looked at simple point mass systems to find the fundamental properties before moving on to the generalised MBS.

To apply the principles introduced in the previous sections to point masses, it is necessary to distinguish between **external** and **internal** forces. External forces are acting on each mass point individually, whereas the internal forces are created by the relation between different point masses like it is shown in Fig. 2.2. This leads to a motion equation for a mass point different from the one m_i as in (2.1)

$$\underbrace{\sum_j \mathbf{F}_{ji}}_{internal} + \underbrace{\mathbf{F}_i^e}_{external\ force} = \dot{\mathbf{p}}_i \qquad (2.16)$$

Apparently, in case of $i = j$ the inner force has to be $\mathbf{F}_{jj} = 0$ meaning that a particle mass exerts no force on itself. By summing up all moments one gets the total moment in which the Newton's second law (2.2) can be inserted to obtain the equation for all mass points.

$$\frac{d^2}{dt^2} \sum_i m_i \mathbf{r}_i = \sum_i F_i^{ext} + \sum_i \sum_{i \neq j} \mathbf{F}_{ij} \qquad (2.17)$$

Fig. 2.2 An example of
force-relationships between
point-masses

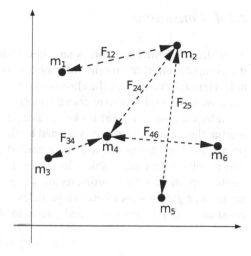

Presuming further the validity of the third law (*actio et reactio*) for the relational
force between the particles, then $\mathbf{F}_{ji} = \mathbf{F}_{ij}$ holds and all inner forces vanish. In
physics, this assumption is also known as *the principle of weak interaction* and is
valid for many materials, especially those which are treated in this work. Similarly,
one can reduce the left hand side of the term by condensing the masses to the centre
of gravity $\mathbf{R}$. This is an equivalent definition obtained by creating the weighted
average.

$$\mathbf{R} = \frac{\sum_i m_i \mathbf{r}_i}{\sum_i m_i} = \frac{\sum_i m_i \mathbf{r}_i}{M} \tag{2.18}$$

Consequently, a simplified motion equation is obtained for the centre of masses.

$$M \frac{d^2}{dt^2} \mathbf{R} = \sum_i \mathbf{F}_i^e \tag{2.19}$$

Finally, external forces are solely affecting the centre of masses of a MBS for which
the weak interaction holds. Analogously, one can extend this statement to moments
which yields

$$\mathbf{P} \equiv \sum_i \mathbf{p}_i = M \frac{d\mathbf{R}}{dt} \tag{2.20}$$

From the latter equation it follows that the total moment is preserved if no external
forces are acting on the MBS.

   Coming now to the work done within a MBS: by naturally treating the total work
as the sum of the work done by all mass points. The work of (2.7) is adopted and
extended by the force relation. This leads to an expression for the total work

$$W_{ab} = \sum_{i \neq j} \int_a^b \mathbf{F}_{ji} \cdot d\mathbf{s}_i + \sum_i \int_a^b F_i^{ext} \cdot d\mathbf{s}_i \tag{2.21}$$

2.1.4 Constraints

After deducing the inner forces and their influence on the total motion of a MBS, the computation of the motion of a system is possible by the formulas introduced beforehand. Nevertheless, the class of systems which can be simulated is quite small since we have no ability to restrict the motion to an allowable space. As an example, consider the case of a pearl necklace. The pearls not only have to follow the law of inertia, their motion is also constrained by the chain that might be fastened to some points. Such constraints have to be regarded in the computations. In this example, two problems become visible which have to be considered appropriately. First, it might happen that the coordinates of some or all n particles are not independent anymore, e.g the distance between particles representing a rigid body have to remain constant. Formally, particles i and j have to obey the condition

$$\mathbf{r}_i^2 - \mathbf{r}_j^2 = l_{ij}^2 \qquad (2.22)$$

Thus the coordinates are related to each other by the constraint and the motion of each particle in space cannot be solved individually. With the introduction of so-called **generalised coordinates** q_i one can circumvent the problem. The generalised coordinates are used to derive the $3n$ particle coordinates by employing functions of q_i

$$\mathbf{r}_i = \mathbf{r}_i(q_1, q_2, \ldots, q_{3N-k}, t) \quad \forall i = 1, 2, \ldots, N \qquad (2.23)$$

where the generalised coordinates are not necessarily coinciding with spatial dimensions. Depending on the problem and its constraints, a transformation, e.g. expressing particle positions in polar coordinates, could be beneficial in solving the motion of the entire system. In general, there are different types of constraints that can be categorised. The first and easiest to handle is the **holonomic** (*Greek: holistic, integrable*) constraint which can be expressed by functions of the form

$$f(\mathbf{r}_i, t) = 0$$

One example of this type has been given by (2.22) for a rigid body. Its solution can be found by the general coordinates. Constraints that cannot be expressed by such functions and are therefore not directly dependent on positions are consequently called **nonholonomic**. These typically require to solve a differential equation in the first place. An example for such a constraint would be a rolling disk on a plane, where the movement of centre of the disk depends on the velocity at the stationary contact point (Fig. 2.3). Further categorisation is being made by time-dependency, namely **rheonomous** (*Greek, in flow*) for variable in time and **scleronomous** for time-independent constraints.

Besides the dependency of coordinates, another problem becomes visible when constraints are given for a motion. The forces induced by the constraints are not *a priori* known. Therefore, it cannot simply the equation of motion be solved as these forces are only reactions upon the movement. But it is known that the constraint forces are holding the system within the allowable range, respectively within

Fig. 2.3 Rolling disc on a
curved path with gravity

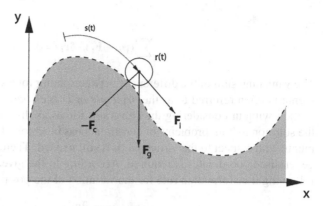

the **configuration space** of our system, being the set of all admissible coordinates
$\mathbf{r}_i$ aggregated to tuples in $\mathbb{R}^{3N}$.

It is further a (compatible) **virtual displacement** defined as an infinitesimal
change in the configuration namely in the positions (or generalised coordinates)
of the system that is consistent with the constraint equations at a fixed time t, i.e.
displacements are perpendicular to the constraints. The displacement $\delta\mathbf{r}$ is called
virtual to distinguish it from a real displacement caused by the change of (*dynamical*) forces and constraints within a time step dt.

If it is supposed having a system which resides within a state of equilibrium such
that the net force $\mathbf{F}_i = 0$ vanishes on all particles i, and when further the virtual
displacement is added to the system, then the positional change $\delta\mathbf{r}$ does no work as
the system is in an equilibrium. Therefore, one can write for the work

$$\sum_i \mathbf{F}_i \cdot \delta\mathbf{r} = 0 \qquad (2.24)$$

Equation (2.24) is known as the **principle of virtual work** which implies that virtual
displacements $\delta\mathbf{r}$ on systems in equilibrium do no virtual work.

The force $\mathbf{F}_i$ can also be written as the sum of dynamical (external) and constraint
force:

$$\mathbf{F}_i = \mathbf{F}_i^{(d)} + \mathbf{F}_i^{(c)} \qquad (2.25)$$

Together with a virtual displacement, one gets for the entire system

$$\sum_i \mathbf{F}_i^{(d)} \cdot \delta\mathbf{r}_i + \sum_i \mathbf{F}_i^{(c)} \cdot \delta\mathbf{r}_i = 0 \qquad (2.26)$$

As consequence of the virtual work principle it can be seen that the right term has to
be zero as the allowed virtual displacements are orthogonal to the constraint forces.
Therefore the separation of the net force in the dynamic and constraint component
can be omitted. Moreover, since the system is in equilibrium state, it is also known
as a consequence of the previous observations on Newton's 2nd law of dynamics
that these forces are then equal to the counteracting forces created by $\dot{\mathbf{p}}_i$. Resulting

in

$$\sum_i (\dot{\mathbf{p}}_i - \mathbf{F}_i) \cdot \delta\mathbf{r}_i = 0. \tag{2.27}$$

The vanishing sum of the differences between acting forces and time-derivative momenta is often referred to as the *principle of D'Alembert*. It allows to compute the motion without considering the constraint forces as they are eliminated. Although the solution to both problems of constraints has been found, a valid virtual displacements with respect to the virtual work is still needed. Therefore, a transformation in generalised coordinates is required. According to the given mapping in (2.23), the following transformation rule for the variation of the positions

$$\delta\mathbf{r}_i = \sum_j \frac{\partial\mathbf{r}_i}{\partial q_j} \cdot \delta q_j \tag{2.28}$$

is obtained. Inserting into (2.27) leads to

$$\sum_i \sum_j (\dot{\mathbf{p}}_i - \mathbf{F}_i) \cdot \frac{\partial\mathbf{r}_i}{\partial q_j} \cdot \delta q_j = 0 \tag{2.29}$$

The terms of $\dot{\mathbf{p}}_i$ and $\mathbf{F}_i$ are split up to have a closer look at the individual components. The latter term can be used to define a **generalised force** being

$$Q_j = \sum_i \mathbf{F}_i \cdot \frac{\partial\mathbf{r}_i}{\partial q_j} \tag{2.30}$$

Note that a generalised force is a dimensionless quantity.

2.2 Continuum Formulation

In the previous chapters the essential physical equations that govern the motion of bodies in classical mechanics have been defined. It was always assumed that the bodies are rigid and treated as idealistic point masses. But these equations cannot describe the behaviour of bodies with spatial dimension being furthermore deformable. This is the point where the continuum formulation comes in.

First of all, one has to define what continuum actually means. Looking at an object in at the atomic level, the object typically appears as a conglomerate of atoms enforced to stay in a lattice structure with gaps in it. But from a macroscopic point of view, there are no gaps visible and object's material is perceived to be continuous. When it is discussed about large scale behaviour of an object, the material is treated as a continuum and associate with each material point a spatial coordinate related to a physical quantity. As consequence, there are no gaps between those material points and one can compute the derivatives of these quantities as the limit of the medium at the specified point.

For example, obtaining the density of an object at a point $\mathbf{x}$, the volume V can be shrinked to an infinitesimal small region and, assuming the existence of the limit of mass occupying the volume element and the volume itself, leading to

$$\lim_{\Delta V \to 0} \frac{\Delta m}{\Delta V} = \rho(\mathbf{x})$$

As result, the physical laws that were introduced in the case of point masses to the continuum can be applied. Nevertheless, the laws are not completely describing the behaviour of the continuum, since the equations are only accounting for single points. Therefore geometric deformations and material response due to the change in the continuum are not considered by these formulas. For completeness, it is needed to address these properties by adding equations defining the geometric deformation with respect to the undeformed state.

2.2.1 Internal Strains

Given the reference configuration for the body by Ω being a manifold in $\mathbb{R}^3$ with boundary and define a twice continuously differentiable mapping

$$\mathbf{x} : \begin{cases} \Omega \times \mathbb{R} \to \mathbb{R}^3 \\ \mathbf{X}, \ t \mapsto \mathbf{x}(\mathbf{X}, t) \end{cases} \tag{2.31}$$

onto the deformed state $\mathbf{x}$ of the body. Under the assumption of having no gaps and no overlaps (i.e. no self-intersections), it can be concluded that $\mathbf{x}$ is continuous and the mapping has to be bijective as the deformation cannot degenerate. The locations denoted $\mathbf{X}$ are often referred to as **material coordinates**.

With such a mapping the degree of deformation the body has undergone can be estimated. The deformation quantities, the internal strains, create corresponding internal stresses opposing the deformation. The internal strain at point $\mathbf{x}$ with the infinitesimal change of length ds is identified by the mapping (deformation). Observing the squared length, one gets

$$(ds)^2 = (d\mathbf{x}(\mathbf{X}, t))^2 = \sum_{i=1}^{3} \sum_{j=1}^{3} \left\langle \frac{\partial}{\partial X_i} \mathbf{x}(\mathbf{X}, t), \frac{\partial}{\partial X_j} \mathbf{x}(\mathbf{X}, t) \right\rangle dX_i \cdot dX_j \tag{2.32}$$

In order to reflect the strain with respect to the reference configuration, the reference length denoted by dS has to be consider, thus it follows

$$(ds)^2 - (dS)^2 = \sum_{i=1}^{3} \sum_{j=1}^{3} \left(\left\langle \frac{\partial}{\partial X_i} \mathbf{x}(\mathbf{X}, t), \frac{\partial}{\partial X_j} \mathbf{x}(\mathbf{X}, t) \right\rangle - \delta_{ij} \right) dX_i \cdot dX_j \tag{2.33}$$

whereas δ_{ij} is the *kronecker-delta*, being one **iff** $i = j$ and zero otherwise. Substituting the term on the right hand side with

$$(ds)^2 - (dS)^2 \equiv 2\varepsilon_{ij}dX_i \cdot dX_j \tag{2.34}$$

To omit the summation over the repeated indices, the *Einstein summation convention* is introduced. The ε_{ij} are the components of **Green's strain tensor** ε^G at a point **x** given by

$$\varepsilon_{ij} = \frac{1}{2}\left(\frac{\partial x_k}{\partial X_i}\frac{\partial x_k}{\partial X_j} - \delta_{ij}\right) \tag{2.35}$$

Remark: The strain tensor is strongly related to the metric tensor. For each vector pair of the embedding space the metric tensor creates a positive definite, symmetric bi-linear form with which one can measure lengths or angles in Riemannian manifolds. Here, we measure length changes under our mapping $\mathbf{x}(\mathbf{X}, t)$ which can be interpreted as a local deformation of an manifold in the Euclidean space $\mathbb{R}^3$. Moreover, the strain tensor is by definition symmetric.

Another notation of the strains is typically created by expressing the deformation in displacements u_i with respect to the reference configuration. Consequently, we can write for the position

$$x_i = u_i + X_i, \qquad \frac{\partial x_i}{\partial X_j} = \frac{\partial u_i}{\partial X_j} + \delta_{ij} \tag{2.36}$$

Insertion into (2.35) gives

$$\varepsilon_{ij} = \frac{1}{2}\left(\frac{\partial u_i}{\partial X_j} + \frac{\partial u_j}{\partial X_i} + \frac{\partial u_k}{\partial X_i}\frac{\partial u_k}{\partial X_j}\right) \tag{2.37}$$

Components ε_{ii} on the diagonal are an estimate for the elongation in the coordinate directions, whereas the others are the shear components. A diagonalisation of the strain-tensor is possible by computing the eigenvectors with its associated eigenvalues. The eigenvectors determine the directions of the principal strains describing the volume or surface change. Therefore, the mean volume change is estimated by $\frac{1}{3}\mathrm{tr}(\varepsilon^G)$.

The latter tensor can be further linearised by omitting the nonlinear terms. As a result we obtain **Cauchy's-strain** tensor **C** with its components

$$C_{ij} = \frac{1}{2}\left(\frac{\partial u_i}{\partial X_j} + \frac{\partial u_j}{\partial X_i}\right) \tag{2.38}$$

The benefit of Cauchy's tensor is that it can be computed more efficiently due to its linearity. But at the same time the simplification induces errors for large strains so it should only be used for small strains. Moreover, it does not account for rotations of the frame of reference and is thus variant under rotations. Although the tensor exhibits some drawbacks, its simplicity in computation mostly outweighs the aforementioned problems. With a special treatment of the situation the drawbacks are kept minimal. As the tensor depends on the orientation of the local frame of reference, the local deformation is separated into two transformations. The first is determined by the rigid rotation of the frame of reference and the other by the deformations given by the Cauchy tensor. This approach of separating the rotations from the strain determination is referred to as **co-rotational formulation** and is often favoured for real-time finite element simulation systems [7, 12].

2.2.2 Mechanical Stresses

With the strain tensor a measure for deformation at an arbitrary point within a continuous medium is obtained. As the strains are a result of acting forces, the forces which act on a continuous body has to be described. These fall into the categories of internal and external forces. The latter is further divided into surface forces and body (or volume) forces. Body forces act on the distribution of mass inside the body, e.g. gravity, while surface forces act on the boundary, e.g. contact forces. The former category of internal force is created by the resistance of the material against deformation. Hereby, the force is termed **stress** or **traction** acting on the surface element of an infinitesimal volume dV. Therefore the stress vector $\mathbf{t}$ depends on the surface orientation and yields a pressure (force density) as unit. It is defined by

$$\mathbf{t}(\mathbf{n}) = \lim_{\Delta S \to 0} \frac{\Delta \mathbf{F}(\mathbf{n})}{\Delta S} \tag{2.39}$$

where ΔS denotes the surface element, $\mathbf{n}$ its orientation and $\Delta \mathbf{F}$ the force acting on it. The vector $\mathbf{t} \cdot \mathbf{n}$ is the stress in direction of the surface normal $\mathbf{n}$ and is consequently called **normal stress**. The stress component perpendicular to the normal is called **shear stress**. Since the third law of Newton should also hold for the stresses, it follows that $\mathbf{t}(-\mathbf{n}) = -\mathbf{t}(\mathbf{n})$. In general, the stress vector is not restricted to normal direction. By analysing the dependency between normal and stress vector, one can create a mapping from any surface vector to the resulting stress.

Therefore an infinitesimal tetrahedron as shown in Fig. 2.4 is considered and the stress vectors $\mathbf{t}(\mathbf{e}_i)$ according to the surface elements ΔS_i oriented with respect to the Cartesian basis e_i are denoted. The motion of tetrahedron has to obey the second law of Newton such that all stresses acting on the surfaces are related to the acceleration $\mathbf{a}$ as follows

$$\mathbf{t}(\mathbf{n})\Delta S + \mathbf{t}_1 \Delta S_1 + \mathbf{t}_2 \Delta S_2 + \mathbf{t}_3 \Delta S_3 + \mathbf{F}^{(e)} \rho \Delta V = \mathbf{a} \rho \Delta V \tag{2.40}$$

From a special case of the Gauss divergence theorem, it follows that the total vector area of the closed surface is

$$\Delta S \mathbf{n} - \Delta S_1 \mathbf{e}_1 - \Delta S_2 \mathbf{e}_2 - \Delta S_3 \mathbf{e}_3 = 0 \tag{2.41}$$

Thus it can be written for the surface elements

$$\Delta S_i = (\mathbf{n} \cdot \mathbf{e}_i)\Delta S \quad \text{with } i \in \{1, 2, 3\} \tag{2.42}$$

Fig. 2.4 Tetrahedron with
surface stresses in Cartesian
coordinates

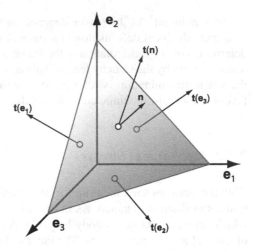

The volume ΔV of the element is computed by the formula

$$\Delta V = \frac{\Delta h}{3} \Delta S \qquad (2.43)$$

where Δh denotes the distance between the origin and the slant face. Substituting
(2.41) and (2.42) in the initial formula (2.40) of the acting forces and dividing by
ΔS leads to

$$\mathbf{t} = \sum_{i=1}^{3} (\mathbf{n} \cdot \mathbf{e}_n) + \rho \frac{\Delta h}{3} (\mathbf{a} - \mathbf{f}) \qquad (2.44)$$

Shrinking the tetrahedron by $\Delta h \to 0$ yields the limit

$$\mathbf{t} = \sum_{i=1}^{3} (\mathbf{n} \cdot \mathbf{e}_i) \mathbf{t}_i \qquad (2.45)$$

The result can be rewritten as a tensorial relationship whereas the stress-vectors $\mathbf{t}_i$
are identified with the columns of the stress tensor. Consequently, an equivalent
tensor notation of the latter equation is obtained if $\mathbf{n}$ is extracted

$$\mathbf{t} = \mathbf{n} \cdot \sigma \qquad (2.46)$$

$$\sigma \equiv \mathbf{e}_i \cdot \mathbf{t}_i \qquad (2.47)$$

each component of the tensor is given by

$$\sigma_{ij} = \mathbf{t}_i \cdot \mathbf{e}_j \qquad (2.48)$$

This notation is possible as the stress tensor defines a property being independent
of $\mathbf{n}$. Another expression of the stress tensor is given by the separation of the or-

Fig. 2.5 Correspondence of
stress components to surface
elements

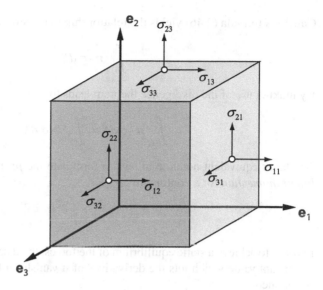

thogonal components of $\mathbf{t}_i$. We can write

$$\mathbf{t}_i = \sigma_{i1}\mathbf{e}_1 + \sigma_{i2}\mathbf{e}_2 + \sigma_{i3}\mathbf{e}_3 \qquad (2.49)$$

$$= \sigma_{ij}\mathbf{e}_j \qquad (2.50)$$

from which it follows that

$$\sigma = \mathbf{e}_j\mathbf{t}_i = \sigma_{ij}\mathbf{e}_i\mathbf{e}_j \qquad (2.51)$$

The entries σ_{ij} represents the force per unit area on an surface element perpendicular to the ith coordinate and the jth coordinate direction (see Fig. 2.5). Equation (2.46) is known as **Cauchy-stress formula** and the tensor σ is termed **Cauchy-stress tensor**. By using the principle of conservation of angular momentum, it can be proven that the tensor needs to be symmetric (cf. [13]).

Similar to the strain tensor, the stress tensor can be diagonalised and the estimation of the eigenvectors yields the principal stresses where the value of diagonal elements $\sigma_{ii} > 0$ indicate tension and $\sigma_{ii} < 0$ indicate pressure. Analogously to the estimation of mean volume change in the strain vector, the mean pressure is given by $\frac{1}{3}\text{tr}(\sigma)$.

The stress tensor σ now allows to describe the internal forces in a body. With $\mathbf{f}$ being the body force per unit mass, it follows for the total body force

$$\int_{\Omega} \rho \mathbf{f} dV \qquad (2.52)$$

Moreover, with $\mathbf{t}$ as surface force per unit area, the traction field of the body is given by

$$\int_{\partial\Omega} \mathbf{t} dS \qquad (2.53)$$

Cauchy's formula (2.46) yields the relationship to the stress tensor

$$\int_{\partial \Omega} \mathbf{n} \cdot \sigma \, dS \tag{2.54}$$

By making use of the divergence theorem leads to

$$\int_{\partial \Omega} \mathbf{n} \cdot \sigma \, dS = \int_{\Omega} \nabla \cdot \sigma \, dV \tag{2.55}$$

The latter equivalent notation allows to formulate the *principle of conservation of linear momentum* for a continuum

$$\nabla \sigma + \rho \mathbf{f} = \rho \frac{\partial^2 \mathbf{u}}{\partial t^2} \tag{2.56}$$

In order to achieve a static equilibrium of the forces, the displacements have to reach a constant value which lets the derivatives of $\mathbf{u}$ vanish. In Cartesian coordinates, it is obtained

$$\frac{\partial \sigma_{ji}}{\partial \mathbf{x}_j} + \rho \mathbf{f}_i = 0 \tag{2.57}$$

2.2.3 Constitutive Equations

The degree of deformation is reflected in the strain tensor and the actual forces acting on a volume element is determined. Next, it is needed to relate the deformations to the forces and thus to define a function describing the response of the material in terms of stresses to the corresponding strains. The so-called **response function** C is a general description of the desired correspondence

$$\sigma = C(\varepsilon) \tag{2.58}$$

The above equation is termed **constitutive equation** and may consider various influences, e.g. thermal conductivity, affecting the response and thus the stresses. If the response of the material is the same at every material coordinate, the material is **homogeneous** otherwise **heterogeneous**. Moreover, if the stress in the material depends on the strain direction, the material is said to be **anisotropic**. If the behaviour differs only in orthogonal strain directions, the material is **orthotropic** (or exhibits orthogonal anisotropy). A material being indifferent to the direction of strain is called **isotropic**.

The theory of linear elasticity considers only ideally elastic materials. A deformed body, under isothermal conditions, is ideally elastic if it recovers its shape completely when external forces are removed. As a consequence, it restricts the response function to a simple one-to-one mapping of the strains to the stresses and C is sufficiently described by a fourth-order tensor with 81 coefficients in total.

This restriction is also known as the **generalised Hooke's Law** which also linearly approximates nonlinear stresses at small strains by the assumption of constant stiffness. Hence, it follows

$$\sigma_{ij} = \sum_{k,l=1}^{3} \mathcal{C}_{ij,kl}\varepsilon_{kl} \qquad (2.59)$$

for the stress components.

Since the strain and stress tensors are symmetric by definition, only six components remain free. A reduction of the coefficients is introduced by the **Kelvin-Voigt** notation leading to the single index notation:

$$\sigma_1 = \sigma_{11}, \quad \sigma_1 = \sigma_{22}, \quad \sigma_1 = \sigma_{33}, \quad \sigma_4 = \sigma_{23}, \quad \sigma_5 = \sigma_{13}, \quad \sigma_6 = \sigma_{12} \qquad (2.60)$$

$$\varepsilon_1 = \varepsilon_{11}, \quad \varepsilon_1 = \varepsilon_{22}, \quad \varepsilon_1 = \varepsilon_{33}, \quad \varepsilon_4 = 2\varepsilon_{23}, \quad \varepsilon_4 = 2\varepsilon_{13}, \quad \varepsilon_4 = 2\varepsilon_{12} \qquad (2.61)$$

The notation now yields the matrix $C \in \mathbb{R}^{6 \times 6}$

$$\begin{pmatrix} \sigma_1 \\ \sigma_2 \\ \sigma_3 \\ \sigma_4 \\ \sigma_5 \\ \sigma_6 \end{pmatrix} = \begin{pmatrix} C_{11} & C_{12} & C_{13} & C_{14} & C_{15} & C_{16} \\ C_{21} & C_{22} & C_{23} & C_{24} & C_{25} & C_{26} \\ C_{31} & C_{32} & C_{33} & C_{34} & C_{35} & C_{36} \\ C_{41} & C_{42} & C_{43} & C_{44} & C_{45} & C_{46} \\ C_{51} & C_{52} & C_{63} & C_{54} & C_{55} & C_{56} \\ C_{61} & C_{62} & C_{73} & C_{64} & C_{65} & C_{66} \end{pmatrix} \begin{pmatrix} \varepsilon_1 \\ \varepsilon_2 \\ \varepsilon_3 \\ \varepsilon_4 \\ \varepsilon_5 \\ \varepsilon_6 \end{pmatrix} \qquad (2.62)$$

Additional simplification is employed by having an idealised relationship in the work done with respect to the strain. A so-called **hyperelastic** material does not absorb the energy that is induced by the strains and releases the energy completely at recovery.

For such hyperelastic materials, the number of free coefficients of $\mathcal{C}$ is reduced to 21. Additional orthotropy or isotropy reduces the number further to 9 or 2 coefficients respectively.

The two remaining free coefficients denoted by λ and μ of isotropic materials are named **Lamé-constants**. These constants can be indirectly obtained by experimentally measuring the engineering parameters of **Young** or **elastic modulus** E and **Poisson's ratio** v or transverse contraction. The latter opposes the material compression and ranges from $-1 < v < \frac{1}{2}$. The former shows the resistance force according to the tension. The Lamé-constants are then deduced from the previously measured parameters with

$$\lambda = \frac{Ev}{(1+v)(1-2v)} \quad \text{and} \quad \mu = \frac{E}{2(1+v)} \qquad (2.63)$$

A special type of linear, isotropic materials (being nonlinear in geometric configuration with Green's tensor) are the **St. Venant-Kirchhoff** materials with the relation

$$\sigma = \lambda \operatorname{tr}(\varepsilon)\mathbf{I} + 2\mu\varepsilon \qquad (2.64)$$

Table 2.1 Mechanical properties of elastic materials

Mechanical property	Symbol	Characteristic	Unit
Elastic modulus	E	Linear slope of σ	N/mm^2
Poisson ratio	ν	Ratio of transverse contraction strain	–
Shear modulus	G	Resistance against shearing	N/mm^2
Bulk modulus	K	Resistance against compression	N/mm^2

and their hyperelastic energy function is given by

$$V = \lambda/2\mathrm{tr}(\varepsilon)^2 + \mu\varepsilon^2 \tag{2.65}$$

By modelling the stress-strain relationship with the St. Venant-Kirchhoff material and the generalised Hooke's law, the constitutive equation becomes very simple. The symmetric matrix is given by

$$C = \begin{pmatrix} 2\mu + \lambda & \lambda & \lambda & 0 & 0 & 0 \\ \lambda & 2\mu + \lambda & \lambda & 0 & 0 & 0 \\ \lambda & \lambda & 2\mu + \lambda & 0 & 0 & 0 \\ 0 & 0 & 0 & 2\mu & 0 & 0 \\ 0 & 0 & 0 & 0 & 2\mu & 0 \\ 0 & 0 & 0 & 0 & 0 & 2\mu \end{pmatrix} \tag{2.66}$$

An additional factor affecting the material behaviour is the dependence of the stress response on the strain rates $\dot{\varepsilon}$. Such behaviour is justified by the fact that materials have decreasing stresses under constant strains (**stress relaxation**). The latter behaviour is also known as **creep**. Another behaviour of some materials in situations of periodic load and unload can be experimentally observed. Here, the anomaly is that the stresses lag behind the strains. The presence of one of these effects classifies the material to be **visco-elastic**. Models considering such phase-shift behaviour are described in the impulse response of the material (in stresses) with respect to a strain unit pulse. A history of strains is required to perform the frequency decomposition of the strain signal yielding the temporal stress response. Therefore, proper modelling demands a considerable amount of computation time and storage which makes a precise model not yet suitable for real-time simulation. The interested reader might look into [10] for a comprehensive view. Nevertheless, ignoring the deformation history still allows to simulate some aspects of such materials.

2.2.4 Energy Principles and Variational Approach

As illustrated in Sect. 2.1.2, quantities of the energy and the work done define the motion of a dynamical system which obeys the principle of least action. In the following the estimation of these quantities will be extended to continuous bodies.

For a continuum, the total work done is contributed by the external W_e and internal W_i work. The former contribution is given by a distributed force field $\mathbf{f}(\mathbf{x})$ acting on the continuum (per unit volume) with its displacement field $\mathbf{u}(\mathbf{x})$. Thus work done by this force field equals

$$W_e = -\int_\Omega \mathbf{f} \cdot \mathbf{u} \, dV \qquad (2.67)$$

assuming the external forces or moments being independent of the actual displacement. The external work W_e equals the potential energy and its negative sign indicates that the work is performed *on* the body.

Besides the external work internal work is exoressed by the integration of stresses as function of the strains. Firstly, the internal work is given

$$U_0 = \int_0^{\varepsilon_{ij}} \sigma_{ij} \, ds \qquad (2.68)$$

at unit element with an actual strain ε_{ij}. The quantity is also often referred to as **strain energy density**. With the existence of such a scalar function U the stresses are said to satisfy the energy equation and are conservative. This implies that the work done depends only on the initial and the final strain. Differentiation of the function leads to

$$\sigma_{ij} = \frac{\partial U_0}{\partial \varepsilon_{ij}} \qquad (2.69)$$

The existence of U_0 is often assumed, e.g. for hyperelastic materials, even under large deformations and nonlinear elasticity. Finally, to obtain the total work done in the body integration over the whole domain Ω is needed to get

$$W_i = U = \int_\Omega U_0 \, dV \qquad (2.70)$$

Furthermore, it can be concluded for a body being in equilibrium under consideration of D'Alembert's principle as in (2.27) that

$$W_i + W_e = \delta W_i + \delta W_e = 0 \qquad (2.71)$$

holds true for the system, where $\delta W = \mathbf{F}\delta u$ denote the work done by actual forces in moving through virtual displacements δu.

A body Ω which is not only subject to a distributed body force $\mathbf{f}(x)$ as before but also a traction field $\mathbf{t}(s)$, yields the total energy of the system regarding virtual displacements

$$\delta W_i + \delta W_e = \int_\Omega \sigma_{ij}\delta\varepsilon_{ij} \, dV - \left(\int_\Omega \delta\mathbf{u} \, dV + \int_{\partial\Omega} \mathbf{t} \cdot \delta\mathbf{u} \, dS \right) = 0 \qquad (2.72)$$

The total energy of the system will be denoted by

$$\Pi(\mathbf{u}) = W_i(\mathbf{u}) + W_e(\mathbf{u}) \qquad (2.73)$$

Consequently, a description of the energy residing in a body has been found. With the variation of the displacement field, it is now tied in with the Hamilton principle of least action to determine the motion or configuration respectively according to the formula (2.72).

2.2.4.1 Hamilton's Principle

When one recalls the procedure of Hamilton of varying the tentative path of a particle such that the time-integral of the difference in kinetic and potential energy reaches a minimum, one will see that the procedure is applicable here as well. Hence, it expresses the body forces in terms of kinetic and potential energy in order to find the displacement which leads to a minimum of the aforementioned time integral.

Let the path of each portion of the body be defined as function $\mathbf{u}(\mathbf{x}, t)$ within a time interval between t_1 and t_2. The variation of the path by a virtual displacement needs to satisfy the following condition

$$\delta\mathbf{u}(\mathbf{x}, t_1) = \delta\mathbf{u}(\mathbf{x}, t_2) = 0 \quad \text{for all } \mathbf{x} \tag{2.74}$$

The condition is crucial to regard the tentative path passing the start- and endpoint in addition to any admissible variation. Furthermore, from the virtual displacement it yields

$$\delta\Pi = \int_\Omega \mathbf{f} \cdot \delta\mathbf{u}\, dV - \int_{\partial\Omega} \mathbf{t} \cdot \delta\mathbf{u}\, dS - \int_\Omega \sigma : \delta\varepsilon\, dV \tag{2.75}$$

for the work Π done on the body at time t $\delta\mathbf{u}$.

The formula departs from (2.72) in sign change which is reasoned by the observation of the work done on the body and not in the system. The operator: denotes the dyadic product between tensors defined by

$$\sigma : \varepsilon = \sigma_{ij}\varepsilon_{ij} \tag{2.76}$$

Considering further the work done by change of momentum through the variation of $\mathbf{u}$, one obtains

$$\int_{t_1}^{t_2} \left(\int_\Omega \rho\frac{\partial^2\mathbf{u}}{\partial t^2} \cdot \delta\mathbf{u}\, dV - \left[\int_\Omega (\mathbf{f}\cdot\delta\mathbf{u} - \sigma : \delta\varepsilon)\, dV + \int_{\partial\Omega} (\mathbf{t}\cdot\delta\mathbf{u})\, dS \right] \right) dt = 0. \tag{2.77}$$

With partial integration and using the boundary conditions of $\delta\mathbf{u}$ it leads to

$$-\int_{t_1}^{t_2} \left(\underbrace{\int_\Omega \rho\frac{\partial\mathbf{u}}{\partial t} \cdot \frac{\partial\delta\mathbf{u}}{\partial t}\, dV}_{\text{kinetic energy}:=T} + \underbrace{\int_\Omega (\mathbf{f}\cdot\delta\mathbf{u} - \sigma : \delta\varepsilon)\, dV + \int_{\partial\Omega} (\mathbf{t}\cdot\delta\mathbf{u})\, dS}_{\text{potential energy}:=\Pi} \right) dt = 0$$

$$\tag{2.78}$$

Thus, the terms yield the Hamilton's Principle of least action mathematically expressed by

$$\delta \int_{t_1}^{t_2} (T - V)\, dt = 0. \tag{2.79}$$

With the transformation into the latter principle, the solution in terms of displacements is found w.r.t. the forces **f** and **t** one may apply. These may arise when employing gravitational and interaction (constraint) forces, respectively. The forces on the border induced by a traction field are related to the contact problem which will be discussed in Chap. 3.

2.3 Numerical Simulation

With the equations derived so far it is possible to describe and compute the mechanics of a continuum, but the positional variation in the object in terms of displacements requires to find a finite subspace of functions defined on the domain fulfilling the boundary conditions of the problem. To find a solution to a problem with complex domain would be impractical or even impossible. It is therefore not only convenient but also necessary to have rather an approximate than the exact solution.

2.3.1 Discretisation and Solution

The idea in finding a numerical solution to the stated variational problem is to reduce the solution function space V to finite dimensional subspace V_0 with dimension n. The so-called Ritz method yields an approximation U_n composed by elements φ_i of basis of the subspace V_o and appropriately chosen coefficients c_i:

$$\mathbf{u}(\mathbf{x}) \approx \mathbf{U}_n(x) = \sum_{i=1}^{n} c_i \varphi_i(\mathbf{x}) \tag{2.80}$$

If $\Pi(\mathbf{u})$ defines the functional and the approximation is of interest then **u** has to be substituted with $\mathbf{U}_n$ such that the variational problem of $\delta \Pi = 0$ becomes

$$\delta \Pi = \frac{\partial \Pi}{\partial c_1} \delta c_1 + \cdots + \frac{\partial \Pi}{\partial c_n} \delta c_n \tag{2.81}$$

From the definition of basis elements it follows that the derivatives are linearly independent

$$\frac{\partial \Pi}{\partial c_i} \delta c_i = 0, \quad \forall i = 1, \dots, n \tag{2.82}$$

Another method in reducing the problem is the transformation into a so-called **weak form**. Since the solution space was reduced by the subspace elements, the solution of the initial problem might not lie in the approximation space. Here the weak form comes into play by reformulating the problem in an alleviated form. Instead of enforcing the approximation solve the exact differential equation of $\delta \Pi = 0$, it has just to satisfy the equation in weighted integral

$$\int_0^1 w(\mathbf{x}) \underbrace{\delta \Pi (\mathbf{U}_n)}_{\equiv R} \, dx = 0 \qquad (2.83)$$

where $w(\mathbf{x})$ is a weight function and $\delta \Pi (\mathbf{U}_n) \neq 0$ the residual R in the differential equation. The new formulation is equivalent to the original problem if (2.83) is satisfied for all suitable $w(\mathbf{x})$ and sufficiently smooth solution.

To define an appropriate subspace, the original domain Ω is decomposed into a finite sub-domain Ω_e which are usually represented by triangles or tetrahedrons depending on the dimensionality of the manifold. These elements simplify the finding of appropriate basis functions to fulfil the needs of the Ritz method. The decomposition of the domain of problem into elements with the definition of appropriate weight functions is called **Finite Element Method**. Each element of the sub-domain is transformed to the reference element which is used here to solve the equation on a simple domain. The weight functions within each element are represented by interpolation functions forming the subspace basis. Those functions interpolate between nodes the assigned value of the approximated function.

In the following the problem will be formulated on such an element. For simplicity, it will be observed on a tetrahedral element of unit length in Cartesian coordinates. Further, the support is chosen by linear interpolation functions

$$\psi_1 = 1 - \zeta - \nu - \xi \qquad (2.84)$$

$$\psi_2 = \zeta \qquad (2.85)$$

$$\psi_3 = \nu \qquad (2.86)$$

$$\psi_4 = \xi \qquad (2.87)$$

to describe a position $\mathbf{x}$ in an element by local coordinates $\zeta, \nu, \xi \in [0, 1]$. The interpolation condition ensures continuity of the function across the borders of elements which requires that

$$\psi_i^e(\mathbf{x}_j^e) = \delta_{ij} \qquad (2.88)$$

is satisfied, where $\mathbf{x}_j^e$ is the position of the j-th node in the e-th element.

The application of the Ritz method to a tetrahedral subspace element Ω_e with four nodes leads to

$$\mathbf{u}^e(\mathbf{x}) = \sum_{i=1}^4 c_i^e \varphi_i^e(\mathbf{x}) = \sum_{i=1}^4 \mathbf{u}_i^e \psi_i^e(\mathbf{x}) \qquad (2.89)$$

locally approximating the displacement **u**. Thus, it can be written for **u** and δ**u** in matrix form by the use of interpolation functions

$$\mathbf{u} = \begin{pmatrix} u_x \\ u_y \\ u_z \end{pmatrix} = \psi^e \Delta^e, \qquad \delta\mathbf{u} = \begin{pmatrix} \delta u_x \\ \delta u_y \\ \delta u_z \end{pmatrix} = \psi^e \delta \Delta^e \qquad (2.90)$$

where

$$\psi^e = \begin{bmatrix} \psi_1 & 0 & 0 & \psi_2 & 0 & 0 & \cdots & \psi_n & 0 & 0 \\ 0 & \psi_1 & 0 & 0 & \psi_2 & 0 & 0 & \cdots & \psi_n & 0 \\ 0 & 0 & \psi_1 & 0 & 0 & \psi_2 & 0 & 0 & \cdots & \psi_n \end{bmatrix} \qquad (2.91)$$

$$\Delta^e = \begin{bmatrix} u_x^1 & u_y^1 & u_z^1 & u_x^2 & u_y^2 & u_z^2 & \cdots & u_x^n & u_y^n & u_z^n \end{bmatrix} \qquad (2.92)$$

$$\delta\Delta^e = \begin{bmatrix} \delta u_x^1 & \delta u_y^1 & \delta u_z^1 & \delta u_x^2 & \delta u_y^2 & \delta u_z^2 & \cdots & \delta u_x^n & \delta u_y^n & \delta u_z^n \end{bmatrix} \qquad (2.93)$$

To transform the potential Π of (2.78) in vector form on the element, the stress-strain-relationship has to be expressed in Kelvin-Voigt notation and replaced strain by the displacement. Thus, it follows

$$\varepsilon = D\mathbf{u} \qquad (2.94)$$

with

$$\varepsilon = \begin{pmatrix} \varepsilon_{xx} & \varepsilon_{yy} & \varepsilon_{zz} & 2\varepsilon_{xz} & 2\varepsilon_{yz} & 2\varepsilon_{xy} \end{pmatrix}^T \qquad (2.95)$$

$$D^T = \begin{bmatrix} \frac{\partial}{\partial x} & 0 & 0 & \frac{\partial}{\partial z} & 0 & \frac{\partial}{\partial y} \\ 0 & \frac{\partial}{\partial y} & 0 & 0 & \frac{\partial}{\partial z} & \frac{\partial}{\partial x} \\ 0 & 0 & \frac{\partial}{\partial z} & \frac{\partial}{\partial x} & \frac{\partial}{\partial y} & 0 \end{bmatrix}. \qquad (2.96)$$

The matrix form is obtained for the motion equation (2.56) of Newton's second law

$$\Delta^T \sigma + \mathbf{f} = \rho \frac{\partial^2 \mathbf{u}}{\partial^2 t} \qquad (2.97)$$

Thus it results from the principle of the virtual displacement as in (2.72) and the strain-stress-relation (2.62) on the reference element, that

$$\int_{\Omega^e} \underbrace{\left[(D\delta\mathbf{u})^T C(D\mathbf{u}) + \rho \delta\mathbf{u}^T \frac{\partial^2 \mathbf{u}}{\partial t^2} \right]}_{=K^e \mathbf{u}^e + M^e \ddot{\mathbf{u}}^e} dV - \int_{\Omega^e} \underbrace{(\delta\mathbf{u})^T \mathbf{f}}_{=\mathbf{f}^e} dV - \int_{\partial\Omega^e} \underbrace{(\delta\mathbf{u})^T \mathbf{t}}_{=Q^e} dS = 0$$

$$(2.98)$$

where each integrand is evaluated with respect to the subspace basis $\langle \psi_i \rangle$. Since (2.98) must hold under all admissible virtual displacements, the matrices can be computed and the coefficients assembled for $\mathbf{u}^e$ in one matrix given by

$$K_{ij}^e = \int_{\Omega^e} (D\psi_i)^T C (D\psi_j) dV \qquad \text{(stiffness matrix)}$$

$$M_{ij}^e = \int_{\Omega^e} \rho \psi_i^T \psi_j dV \qquad \text{(mass matrix)}$$

$$f_i^e = \int_{\Omega^e} \psi_i^T \mathbf{f} dV, \quad Q_i^e = \int_{\partial\Omega^e} \psi_i^T \mathbf{t} dS, \quad \mathbf{F}^e := \mathbf{f}_i^e + \mathbf{Q}^e \quad \text{(external force vector)}$$

Considering further the time-dependency of $\mathbf{u}$, the evaluation yields a dynamical system

$$M^e \ddot{\mathbf{u}}(t) + \dot{\mathbf{u}}(t) + K^e \mathbf{u} = \mathbf{F}^e \qquad (2.99)$$

To find the global solution to the problem, it is needed bring the nodes of the elements into correspondence. With the identification of element nodes u_i^e with a global node, a global linear system is obtained. Here, the local value of the node u_i^e is substituted by a global value U_m at each element. The complete system follows from the initial variational problem (2.81) and the assembly:

$$\delta\Pi = \sum_{I=1}^{M} \frac{\partial\Pi}{\partial\mathbf{U}_I} \delta\mathbf{U}_I = 0 \qquad (2.100)$$

with M being the number of global nodes and Π the sum of the potential energy functions Π^e of the N elements.

$$\Pi = \sum_{e=1}^{N} \Pi^e \qquad (2.101)$$

2.3.2 Textile Simulation Model

While earlier interactive simulators of textiles were mostly modelled by a linear spring particle system, a great improvement was made by employing a non-linear differential equation system allowing to express the physical laws by a particle system. With the approach of Baraff et al. [2] it is now possible to simulate the non-linear material behaviour in large time steps without loosing numerical stability. This approach is the basis of the numerical computations used in our VR system. In the following the representation of a textile and its simulation will be explained in analogy to the FE method.

Since the basis of the approach is a particle system, i.e. a discretisation of the continuous material, the two dimensional manifold embedded in $\mathbb{R}^3$ describing the textile is condensed into mass points. A triangulation as it shown in Fig. 2.6 preserves the topology of the textile and determines the mechanical relations of the

Fig. 2.6 Representation of a
textile by a particle system
with mass lumping

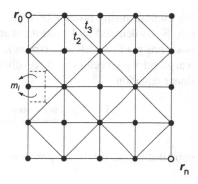

Fig. 2.7 Strains resulting in
triangle deformation

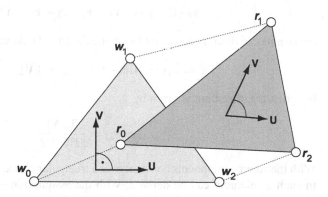

particle system by linear elements. Each triangle represents a part of the original
continuous surface and the associated vertices store the physical quantities, e.g.
energy and inertia. With these quantities, the vertices become particles, i.e. small
masses governed by the equation of motion (2.2). All in all, a particle has a mass
m and a time-dependent position denoted by $\mathbf{x}(t)$ with its time-derivatives as velocity $\dot{\mathbf{x}}(t)$ and acceleration $\ddot{\mathbf{x}}(t)$, respectively. By combining all positions of the n
particles, we get a single vector $\mathbf{r} \in \mathbb{R}^{3n}$.

Now, according to the equation of motion together with the inner forces written
as a potential the following equation for the complete particle system is obtained

$$\ddot{\mathbf{r}} = M^{-1}\left(-\frac{\partial V}{\partial \mathbf{r}} + \mathbf{F}\right) \tag{2.102}$$

where M is the matrix of all particle masses and $\mathbf{F}$ denotes external forces. Within
the generic potential V in (2.102), the material behaviour model is hidden, i.e. it
models the tensile, shear and bend stresses in dependence of the strains. The tensile
and shear stresses inside each triangle corresponding to a fraction of the textile are
determined. To calculate the strains, the particles $\mathbf{r}_i$, $i = 1, \ldots, 3$ of each triangle are
associated with a reference configuration $\mathbf{w}_i$ in a two dimensional space. Figure 2.7
shows a triangle in (reference and deformed) configuration with the local vectors $\mathbf{u}$
and $\mathbf{v}$. The axes of this space correspond to the dominant directions inherent to

the manufacturing process of woven fabrics. The mapping of weft direction vector into $\mathbb{R}^3$ is denoted by $\mathbf{U}$ and the warp direction corresponds to $\mathbf{V}$ respectively. The positions $\mathbf{r}_i$ of the deformed triangle can then be used to calculate the deformed unit warp and weft vectors in three dimensional space for this triangle by solving the linear equation

$$(\mathbf{U} \quad \mathbf{V}) \cdot \underbrace{(\mathbf{w}_2 - \mathbf{w}_1 \quad \mathbf{w}_3 - \mathbf{w}_1)}_{:=A\in\mathbb{R}^{2\times2}} = (\mathbf{r}_2 - \mathbf{r}_1 \quad \mathbf{r}_3 - \mathbf{r}_1) \qquad (2.103)$$

substituting the reference positions with the constant matrix A and inverting leads to

$$\Leftrightarrow (\mathbf{U} \quad \mathbf{V}) = (\mathbf{r}_2 - \mathbf{r}_1 \quad \mathbf{r}_3 - \mathbf{r}_1) \cdot A^{-1} \qquad (2.104)$$

the strains can directly associated with the deformed axis vectors

$$\Rightarrow \varepsilon_{11} = \|\mathbf{U}\| - 1 \qquad \varepsilon_{vv} = \|\mathbf{V}\| - 1 \qquad (2.105)$$

For shearing, the strain is given by

$$\varepsilon_{12} = \varepsilon_{21} = \frac{\langle \mathbf{U}, \mathbf{V}\rangle}{\|\mathbf{U}\|\,\|\mathbf{V}\|} \qquad (2.106)$$

With the strain components constant over a triangular element, the stresses occurring in such an element can be derived. With the assumption of having a conservative energy, the strain energy function is formulated by

$$U_m(\varepsilon) = \int_0^{\varepsilon_m} \sigma_m(s)\,ds \quad \text{with } m \in \{11, 12, 22\} \qquad (2.107)$$

In a triangle element, ε depends on the particle positions. It follows for the work U^e done on a triangle from (2.70)

$$U^e = \frac{|A|}{2} \sum_m \int_0^{\varepsilon_m} \sigma_m(s)\,ds \qquad (2.108)$$

with $\frac{|A|}{2}$ being the area of the triangle in the reference configuration.

As the textile is modelled as a two-dimensional manifold an additional work function U^b has to be defined accounting for the stresses caused by out-of-plane deformations, termed bend forces.

The integration of bend energy within the simulation model, require additional constitutive equations necessary to reflect such stresses. The approach to consider these stress is made by defining out-of-plane strains in terms of curvature. The theory of plates or shells provides different possibilities of representing the bend energy within a plate or (curved) shell, but the in-plane stresses will be considered first. Consequently, the potential defined by

$$V^e := U^e$$

The forces $\mathbf{F}_i$ occurring at each particle can be computed by (2.69) and with ε as function of position $\mathbf{r}_i$, one gets

$$\mathbf{F}_i = -\nabla V^e = -\frac{|A|}{2}\nabla V^e(\varepsilon) \qquad (2.109)$$

$$= -\frac{|A|}{2} \cdot \sum_m \sigma_m(\varepsilon_m)\left(\frac{\partial \varepsilon_m}{\partial \mathbf{r}_i}\right)^T \qquad (2.110)$$

For the numerical time-integration, which will be discussed in the next section, the evolution of force $\mathbf{F}_i$ w.r.t. to the particle position $\mathbf{r}_j$ is required. Therefore, derivation leads to

$$\frac{\partial \mathbf{F}_i}{\partial \mathbf{r}_j} = -\frac{|A|}{2}\sum_n\sum_m \frac{\partial \sigma_m(\varepsilon_m)}{\partial \varepsilon_m}\frac{\partial \varepsilon_n}{\partial \mathbf{r}_j}\frac{\varepsilon_m}{\partial \mathbf{r}_i} \qquad (2.111)$$

$$+ \sum_n \sigma_m(\varepsilon_m)\frac{\partial^2 \varepsilon_n}{\partial \mathbf{P}_i \partial \mathbf{P}_j} \quad \text{with } n \in \{11, 12, 22\} \qquad (2.112)$$

2.3.3 Optimised Force Calculations

Since these calculations are quite expensive in terms of floating point operations, Volino et al. proposed in [15] to easier to compute force calculations the following formulas modelling tensile behaviour.

$$\varepsilon_{11} = \|\mathbf{U}\| - 1 \qquad \varepsilon_{12} = \frac{\|\mathbf{U}+\mathbf{V}\|}{\sqrt{2}} - \frac{\|\mathbf{U}-\mathbf{V}\|}{\sqrt{2}} \qquad (2.113)$$

$$\varepsilon_{22} = \|\mathbf{V}\| - 1 \qquad (2.114)$$

for the strain tensor. The initial formulation differs only for shear strain from (2.106) and (2.105) by being dependent on the length of $\mathbf{U}$ and $\mathbf{V}$. It is justified by having better accuracy at large deformations. For real-time simulation, the strain calculations had to be even further reduced in complexity by

$$\hat{\varepsilon}_{11} = \frac{\mathbf{U}^T\mathbf{U} - 1}{2} \qquad \hat{\varepsilon}_{12} = \mathbf{U}^T\mathbf{V} = \|\mathbf{U}\|\,\|\mathbf{V}\|\cos\angle\mathbf{U}, \mathbf{V} \qquad (2.115)$$

$$\hat{\varepsilon}_{22} = \frac{\mathbf{V}^T\mathbf{V} - 1}{2} \qquad (2.116)$$

With such a simplification, the costly square roots are avoided and the new functions are differentiable everywhere. However, with these formulas one gets a nonlinearity in ε_{11} and ε_{22} whereas in the ε_{12} component linear relation w.r.t. $\|\mathbf{U}\|$ and $\|\mathbf{V}\|$. Therefore, the latter is only applicable if the strain stress functions given by σ are reparameterised accordingly.

$$\int_0^{\varepsilon(x)} \sigma_1(s)ds = \int_0^{\hat{\varepsilon}(x)} \hat{\sigma}(s)ds + C \tag{2.117}$$

$$\Leftrightarrow \quad \hat{\sigma}(\hat{\varepsilon}) = \frac{\partial}{\partial\hat{\varepsilon}} \int_0^{\varepsilon(x)} \sigma(s)ds \tag{2.118}$$

$$= \sigma(\varepsilon(x))\frac{\partial}{\partial\hat{\varepsilon}}\varepsilon(x) \tag{2.119}$$

$$= \sigma(\varepsilon(x))\frac{\partial}{\partial x}\varepsilon(x)\frac{\partial}{\partial\hat{\varepsilon}}x, \quad \text{with } x = \hat{\varepsilon}^{-1}(\hat{\varepsilon}) \tag{2.120}$$

As an example of for remapping from the accurate ε_{11} strain component given by (2.113) to the simple $\hat{\varepsilon}_{11}$ in (2.115), one has to set x to $|\mathbf{U}|$ and thus it follows for the stress value

$$\hat{\sigma}_{11}(\hat{\varepsilon}_{11}) = \frac{\sigma_{11}(\sqrt{2\hat{\varepsilon}_{11}+1}-1)}{\sqrt{2\hat{\varepsilon}_{11}+1}}$$

The calculation of bending forces follows the approach of Volino et al. described in [17]. The basic idea of the computations is to use the discrete differential operators given by the mesh parametrisation to determine the curvature of the underlying smooth surface discretised by the triangles.

Here, the textile treated as a regular surface being the familiar mapping of $\mathbf{S}$: $P \subset \mathbb{R}^2 \to M \subset \mathbb{R}^3$ with the additional property of at least C^2-continuity. At any point $\mathbf{p} \in P$ of the regular surface one can find a curve γ going through the point $\mathbf{p}$ and sharing the same curvature $\frac{\partial^2}{\partial\gamma^2}S(\gamma(p))$ as the surface at this point. Without loss of generality, it is assumed that the coupled mapping of the curve $S(\gamma(s)) : \mathbb{R} \to M$ is arc length parameterised.

Assuming further the textile surface does not vary much in local shear strain, the length of the first derivation of the surface curve $\|\frac{\partial}{\partial t}S(\gamma(t))\|$ is equal in length of the derivative in parameter space $\|\frac{\partial}{\partial t}\gamma(t)\|$. This is supported by the fact that tense threads are generally not curved and thus finding γ satisfying $\|\frac{\partial}{\partial t}\gamma(t)\| = 1$ is comparatively easy.

The easily found discrete differential operators by the aforementioned assumptions are used to compute the second derivatives of the surface in the principal and shear direction from three and four particles respectively. Strain function related to the curvature is then defined by

$$\varepsilon(\gamma, t_0) = \left\| \frac{\partial}{\partial t^2}S(\gamma(t)) \right\|_{t=t_0} \tag{2.121}$$

$$= \left\| \sum_{i=1}^n g_i \mathbf{r}_i \right\| \tag{2.122}$$

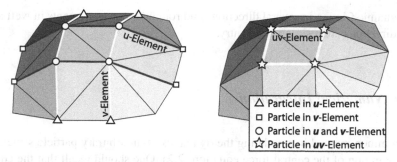

Fig. 2.8 Two types of bending elements (b^3) in *left* and (b^4) in *right figure*

Whereas g_i is the weight of the particle position $\mathbf{r}_i$ in the discrete case. By linearising the stress function to $\sigma(\varepsilon) = B\varepsilon$ we get for the bending force $\mathbf{F}_j$ at particle $\mathbf{r}_j$ a weighted sum of particle positions

$$\mathbf{F}_j = -\frac{\partial}{\partial r_j} V^b(\mathbf{r}) = \frac{\partial}{\partial r_j} \int_0^\varepsilon \sigma(s)ds = \frac{\partial}{\partial r_j} \frac{1}{2} B\varepsilon^2 \qquad (2.123)$$

$$= B \left\| \sum_{i=1}^n g_i \mathbf{r}_i \right\| \frac{\partial}{\partial r_j} \varepsilon \qquad (2.124)$$

$$-\frac{1}{2} B \left\| \sum_{i=1}^n g_i \mathbf{r}_i \right\| \left\| \sum_{i=1}^n g_i \mathbf{r}_i \right\|^{-1} 2 \left(\sum_{i=1}^n g_i \mathbf{r}_i \right) g_j \qquad (2.125)$$

$$= B g_j \sum_{i=1}^n g_i \mathbf{r}_i \qquad (2.126)$$

The exposed interface of the physical simulation library allows to create bending elements associated to the particles with arbitrary weights of g_i and a bending resilience factor B. Figure 2.8 shows the two types of elements which have to be defined on the triangle mesh accounting for bending in **u**-, **v**- and shear direction.

Finally, internal viscosity of the materials in the different contributions of to the force (tensile, shear and bending) are computed by replacing the positional parameter $\mathbf{r}_i$ with its time derivative $\dot{\mathbf{r}}_i$.

In order to provide a variety of virtual fabrics, the simulation model needs retrieve the material parameters from a database containing the information for the visual and haptic rendering. For the latter, each entry features the set of strain-stress functions accounting for the response in each parameter direction. The exact response values are typically obtained by material tests, e.g. tensile test. For internal representation within the computer, the functions are approximated by spline for in-plane and as a constant factor for out-plane stresses. These functions can be dynamically exchanged in the numerical simulation to represent any fabric material. Furthermore, one entry contains the density and the visual appearance of the fabric in a bitmap. For the force feedback and the tactile rendering, coefficients of static and dynamic friction in several directions and roughness measurements in weft and warp directions are residing in an entry.

2.3.4 Numerical Integration

The common method in simulating the dynamics of an arbitrary particle system is the integration of the central force equation (2.2). One should recall that the equation defines the forces acting on the particles w.r.t. their inertia. Hence, with the force known, one can compute the resulting acceleration of each particle. From the mechanical point of view, all important information might be obtained, but for the visualisation and simulation, the resulting change in position is also important. Moreover, as the force on a particle is a function of the positions and velocities of several particles, a system of coupled second-order differential equations has to be solved. In general, the initial state of a particle system at time t_0 is known. The position of the system at the time step t_i is determined by the initial position and its velocity $\mathbf{r}(t)$ with $t_0 \leq t \leq t_i$ by

$$\mathbf{r}(t_i) = \mathbf{r}(t_0) + \int_{t_0}^{t_i} \dot{\mathbf{r}}(t) dt \quad \text{with } \mathbf{r}(t_0) = \mathbf{r}_0. \tag{2.127}$$

In the same way, the velocity at time t_i is determined by the acceleration $\ddot{\mathbf{r}}$ and its initial velocity $\dot{\mathbf{r}}_0$

$$\dot{\mathbf{r}}(t_i) = \dot{\mathbf{r}}(t_0) + \int_{t0}^{t_i} \ddot{\mathbf{r}}(t) dt \quad \text{with } \dot{\mathbf{r}}(t_0) = \dot{\mathbf{r}}_0 \tag{2.128}$$

With the known mass of the particles, the acceleration is computed by (2.102). The latter equations govern the motion of the particles and represent the aforementioned differential equation system in integral form. By combining the positions and velocities into a single vector $\mathbf{y}$, the system is reduced to first order:

$$\mathbf{y}' = \begin{pmatrix} \frac{d}{dt}\mathbf{r}_0 \\ \vdots \\ \frac{d}{dt}\mathbf{r}_n \\ \frac{d}{dt}\dot{\mathbf{r}}_0 \\ \vdots \\ \frac{d}{dt}\dot{\mathbf{r}}_n \end{pmatrix} = \begin{pmatrix} \dot{\mathbf{r}}_0 \\ \vdots \\ \dot{\mathbf{r}}_n \\ \frac{1}{m_0}\mathbf{F}_0(\mathbf{r}_0,\ldots,\mathbf{r}_n,\dot{\mathbf{r}}_0,\ldots,\dot{\mathbf{r}}_n) \\ \vdots \\ \frac{1}{m_n}\mathbf{F}_n(\mathbf{r}_0,\ldots,\mathbf{r}_n,\dot{\mathbf{r}}_0,\ldots,\dot{\mathbf{r}}_n) \end{pmatrix} =: \mathbf{f}(\mathbf{y}) \tag{2.129}$$

As the state of the particle system is known at the start of the application, the latter equation describes an initial value problem (IVP).

2.3.5 Numerical Stability

In general, methods for solving initial value problems are only able to find an approximation to the solution. The quality of such an approximation is called **consistency** and the speed in which the approximation is reached is called **convergency**. These attributes can be associated with the used method. Furthermore, a method is determined to be **stable** if for any pair of initial values in a certain distance, the trajectories stay within a upper bound. The methods more precisely exhibit a region of stability in the complex plane. The plane is being defined by the test equation of Dahlquist or test problem

$$\mathbf{y}' = \lambda \mathbf{y}(t) \quad \text{with } \mathbf{y}(0) = \mathbf{y}_0 \tag{2.130}$$

and λ is a complex number with negative real value. The analytical solution is $\mathbf{y}(t) = \mathbf{y}_0 e^{\lambda t}$. Inserting the used method into the test equation with a fixed timestep Δt, the region of stability is the set of complex numbers $\xi = \lambda \Delta t$ which leads to a monotonic sequence of approximations. If the region contains the halfspace $\{\xi \in \mathbb{C} : Re(\xi) < 0\}$, the method is called **A-stable**. This means that the method is independent of the problem $\mathbf{y}'$. Hauth et al. [8] analysed the numerical stability of the methods presented in the next paragraph.

Numerical methods to solve initial value problems can be categorised by their features. According to the number of used supporting points the methods are separated into **single-step** or **multistep** methods. Single-step methods take only the last solution step into account for the computation of the next time step, whereas multistep methods also use solutions for several previous time steps. The benefit of multistep methods is the higher order of consistency with minor additional computational effort compared to single-step methods. Methods are further characterised by being **explicit** or **implicit**. Explicit methods employ only a priori known values for the calculation of the next approximation. Implicit methods use values being additionally computed within the approximation of the next time step. The latter methods require to solve an equation system determining the relation of the additional data to the previously known. While the extended information increases the order of consistency and convergence, implicit methods suffer in speed because of the extra computations. As the VR system uses single-step methods the consideration will be limited to such methods.

The **explicit Euler**-method is the most simple method to compute the numerical solution for a initial value problem. With the choice of a time step width of $\Delta t > 0$ an approximation is found for the subsequent time steps

$$t_i = t_0 + i \Delta t \quad \text{with } i = 1, 2, 3, \dots \tag{2.131}$$

by the iterative equation of the next state vector

$$\mathbf{y}_{i+1} = \mathbf{y}_i + \Delta t \cdot \mathbf{y}_i' = \mathbf{y}_i + \Delta t \cdot \mathbf{f}(\mathbf{y}_i) \tag{2.132}$$

Unfortunately, the simplicity of the method is bought with a drawback in accuracy and stability, i.e. if $\mathbf{y}_i'$ is subject to large variations, the approximation will quickly diverge from the real solution. Differential equations in which $\mathbf{y}'$ features a high variability with respect to small changes in $\mathbf{y}$ are called **stiff**. Taking into account the stiffness of such systems, the computations of the method need to advance with very small time steps in order to be stable. As textile strain-stress functions are mostly non-linear and of high stiffness (e.g. wild silk reaches 490 N/m at strains of 0.7%), the method would loose its benefit of simplicity by computing too many intermediate steps.

The disadvantage of the explicit methods makes implicit methods more favourable, as their increased stability outweighs the additional computational effort. A variety of implicit methods exist which use sophisticated estimates of the average slope during Δt by calculating intermediate $\mathbf{y}'_{i+\delta}$ for several $t_{i+\delta}$. Some even go as far using a $t_{i+\delta} > t_{i+1}$. A common method among these is the **implicit Euler**-method. Due to its higher order in convergence and consistency it does not suffer in stability from stiff equations. The method differs only from the explicit scheme in its iteration rule

$$\mathbf{y}_{i+1} = \mathbf{y}_i + \Delta t \cdot \mathbf{y}'_{i+1} = \mathbf{y}_i + \Delta t \cdot \mathbf{f}(\mathbf{y}_{i+1}) \tag{2.133}$$

The delicate point here is the use of the a priori unknown state $\mathbf{y}_{i+1}$ for the force computations. Thus an estimation of the evolution of the state is being made by a linearisation at $\mathbf{y}_i$. approximation. A Taylor expansion at $\mathbf{y}_i$ yields the new iteration rule

$$\mathbf{y}_{i+1} = \mathbf{y}_i + \Delta t \left(\mathbf{y}'_i + \frac{\partial}{\partial \mathbf{y}_i} \mathbf{y}'_i \cdot \underbrace{(\mathbf{y}_{i+1} - \mathbf{y}_i)}_{\Delta \mathbf{y}} \right) \tag{2.134}$$

The substitution of the state differences by $\Delta \mathbf{y}$ leads to

$$\Delta \mathbf{y} - \Delta t \frac{\partial}{\partial \mathbf{y}_i} \mathbf{y}'_i \cdot \Delta \mathbf{y} = \Delta t \mathbf{y}'_i \quad \Leftrightarrow \tag{2.135}$$

$$\underbrace{\left(\mathbf{I} - \Delta t \frac{\partial}{\partial \mathbf{y}_i} \mathbf{f}(\mathbf{y}_i) \right)}_{:=M} \cdot \Delta \mathbf{y} = \Delta t \underbrace{\mathbf{f}(\mathbf{y}_i)}_{:=\mathbf{b}} \tag{2.136}$$

The equation with unknown state vector is now transformed into a linear equation system, where the system matrix M and the vector $\mathbf{b}$ are directly given by evaluation. For the solution $\Delta \mathbf{y}$ it is thus necessary to invert the matrix.

Since the matrix size is of $6n \times 6n$ which can be very big depending on the amount of particles, a reduction by half can be made by removing the first $3n$ degrees of freedom (position) of the state vector and solve them explicitly by using the evaluation formula $\mathbf{r}(t_{i+1}) = \Delta t (\dot{\mathbf{r}}(t_i) + \Delta \dot{\mathbf{r}})$. For consistency, the derivative $\mathbf{f}$ then needs to be removed from the positions as well. Despite the removal of the first $3n$ entries, the new derivative $\hat{\mathbf{f}}$ is still dependent on the positions, which has to be considered by splitting the derivation w.r.t. state vector into separate terms.

Finally, it results in

$$\left(\mathbf{I} - \Delta t \frac{\partial}{\partial \dot{\mathbf{r}}} \hat{\mathbf{f}} - \Delta t^2 \frac{\partial}{\partial \mathbf{r}} \hat{\mathbf{f}} \right) \Delta \dot{\mathbf{r}} = \Delta t \left(\hat{\mathbf{f}}(\mathbf{r}, \dot{\mathbf{r}}) + \Delta t \frac{\partial}{\partial \mathbf{r}} \hat{\mathbf{f}} \cdot \dot{\mathbf{r}} \right) \tag{2.137}$$

Yet the matrix still remains too big for standard linear solvers like inversion schemes, e.g. LU-decomposition, to use in the context of real-time simulation. Fortunately, the matrix M exhibits some nice properties which can accelerate the calculations. As the matrix is symmetric and positive definite—a result of the physical formulation—some iterative solvers exist, which can significantly reduce the effort of finding the solution. Additionally, the sparsity of the matrix due to low interdependency of the particles allows further optimisation within such solvers.

Although improved stability of implicit integration against explicit integration has been shown, it introduces numerical damping which affects as well the physical and thus directly the haptic realism as the visual realism. But it is possible to combine implicit and explicit Euler-method to find a balance between stability and physical realism. As a result, we benefit from this combination named **implicit-midpoint** Euler-method in achieving the improved accuracy of implicit integration at the same as the computational speed of explicit scheme. These advantages make the method more favourable than other popular higher-order integration methods like BDF-2 (cf. [1]). In [16] it is further shown that with minor damping the stability can be improved being competitive to more stable schemes and the method is more robust to perturbations of the solution caused by collisions.

In the scheme, the ordinary differential equation system given by the state vector $\mathbf{y}_t$ and its first derivative $\mathbf{y}'$ as defined previously is used to estimate the next time step as follows. The time step Δt is split apart into two steps by a split factor α, named *implicity factor* in [16]. In the first integration step, the implicit Euler would yield for advancing $\alpha \Delta t$ in time

$$\mathbf{y}_{t+\alpha\Delta t} = \mathbf{y}_t + \alpha\Delta t \cdot \mathbf{y}'_{t+\alpha\Delta t} \qquad (2.138)$$

adding for the second half of the total time step an explicit Euler integration leads to

$$\mathbf{y}_{t+\Delta t} = \mathbf{y}_t + \alpha\Delta t \cdot \mathbf{y}'_{t+\alpha\Delta t} + (1-\alpha)\Delta t \cdot \mathbf{y}'_{t+\alpha\Delta t} \quad \Leftrightarrow \qquad (2.139)$$

$$\mathbf{y}_{t+\Delta t} = \mathbf{y}_t + \Delta t \cdot \mathbf{y}'_{t+\alpha\Delta t} \qquad (2.140)$$

The main computational effort in the method as always in implicit schemes is the estimation of $\mathbf{y}'_{t+\alpha\Delta t}$. Since the state's derivative at $\alpha\Delta t$ cannot be directly evaluated due to its dependence on the state itself, a Taylor approximation helps to find

$$\mathbf{y}'_{t+\alpha\Delta t} \approx \mathbf{y}_t + \frac{\partial}{\partial \mathbf{y}}\mathbf{y}'_t(\mathbf{y}_{t+\alpha\Delta t} - \mathbf{y}_t) \qquad (2.141)$$

Now $\mathbf{y}$ has to be approximated at $\alpha\Delta t$ by

$$\mathbf{y}_t \approx \mathbf{y}_{t+\alpha\Delta t} - \alpha\Delta t \cdot \mathbf{y}'_{t+\alpha\Delta t} \qquad (2.142)$$

Substituting $\mathbf{y}_t$ obtained from (2.142) in (2.141) of at the new time step we have

$$\mathbf{y}'_{t+\alpha\Delta t} \approx \mathbf{y}'_t + \alpha\Delta t \cdot \frac{\partial}{\partial \mathbf{y}}\mathbf{y}'_t\mathbf{y}'_{t+\alpha\Delta t} \qquad (2.143)$$

Sorting terms followed by matrix inversion yields

$$\Leftrightarrow \mathbf{y}'_{t+\alpha\Delta t} \approx \left(I - \alpha\Delta t \cdot \frac{\partial}{\partial \mathbf{y}} \mathbf{y}'_t \right)^{-1} \mathbf{y}'_t \qquad (2.144)$$

Finally, inserting the result of $\mathbf{y}'_{t+\alpha\Delta t}$ in (2.140)

$$\Rightarrow \quad \mathbf{y}'_{t+\alpha\Delta t} \approx \mathbf{y}_t + \Delta t \cdot \underbrace{\left(I - \alpha\Delta t \frac{\partial}{\partial \mathbf{y}} \mathbf{y}'_t \right)^{-1}}_{:=M} \mathbf{y}'_t \qquad (2.145)$$

The integration ends up nicely with a formula needing no longer values of $\mathbf{y}$ and $\mathbf{y}'$ of an intermediate time step.

The hardest part in the simulation is now reduced to the inversion of the matrix M introduced by (2.145). But since the matrix might contain several millions of entries, a straight forward inversion using standard decompositions is still not suitable for the real-time constraints. The best approach chosen is again an iterative solution of the linear equation system of M. Provided M being symmetric and positive definite and sufficiently small Δt is determined by the following inequality equation.

$$\alpha\Delta t \mathbf{x}^T \frac{\partial}{\partial \mathbf{y}} \dot{\mathbf{y}}^T \mathbf{x} < \mathbf{x}^T \cdot \mathbf{x}$$

$$\Leftrightarrow \quad \Delta t < \frac{1}{\alpha\lambda_m \frac{\partial}{\partial \mathbf{y}} \mathbf{y}'} \qquad (2.146)$$

Whereas λ_m denotes the largest eigenvalue of M. Despite the requirement of conservative forces yielding a symmetric matrix, the matrix M found in the integration is far from being symmetric:

$$M = \begin{pmatrix} I & -\alpha\Delta t I \\ -\alpha\Delta t \frac{\partial \ddot{\mathbf{r}}}{\partial \mathbf{r}} & -\alpha\Delta t \frac{\partial \ddot{\mathbf{r}}}{\partial \dot{\mathbf{r}}} \end{pmatrix} \qquad (2.147)$$

The splitting of M, suggested by [4], into two terms circumvents the problem to solve the complete matrix. This suggestion is supported by the fact that M has physical contributions by $\mathbf{r}$ and $\dot{\mathbf{r}}$ which are also existent in $\mathbf{y}$. Therefore, a split of $\mathbf{y}$ into these physical contributions yields a symmetric matrix $\hat{M}$ for $\dot{\mathbf{r}}$ with

$$\hat{M} = I - (\alpha\Delta t)^2 \frac{\partial \ddot{\mathbf{r}}}{\partial \mathbf{r}} - \alpha\Delta t \frac{\partial \ddot{\mathbf{r}}}{\partial \dot{\mathbf{r}}} \qquad (2.148)$$

and for the upper part of the matrix M

$$\mathbf{r}_{t+\Delta t} = \mathbf{r}_t + \Delta t \cdot (\dot{\mathbf{r}}_t + \alpha \Delta t \ddot{\mathbf{r}}_{t+\alpha \Delta t}) \tag{2.149}$$

$$\mathbf{P}_{t+\Delta t} = \dot{\mathbf{r}}_t + \Delta t \cdot \ddot{\mathbf{r}}_{t+\alpha \Delta t} \tag{2.150}$$

With the splitting of the linear equation given by M into a new symmetric, positive definite $\hat{M}$ an efficient solution has been found for computation of the complete DES, whereas $\hat{M}$ is an implicit Euler step for the acceleration $\ddot{\mathbf{r}}$ and the other part being an explicit Euler step for the position and velocity.

Another benefit in using an iterative solver here, is its successive approximation to the solution. This allows us to restrict the time spent for computing one time step by setting limits for the maximum number of iterations and the error in the computation. Additional advantage is given by choosing the conjugate gradient (CG) method. This iterative method needs less memory for each iteration compared to others. Contrary to decomposition methods, it does not require the matrix to reside completely in memory. Moreover, due to the speciality of our physical problem, the matrix consists of blocks with 9×9 entries being different from zero where particles have direct influence on each other. These non-zero entries are related to the elements that have been defined by the force functions in Sect. 2.3.3. But since the elements have a very small stencil on the textile mesh, the matrix is very sparse. At an iteration step of the CG method one can even more reduce the memory consumption by computing the blocks when they are needed to for the multiplication with the search vector. Having such small memory requirements can drastically reduce the time for the computations due to the ability of local bookkeeping of the main computer's CPU.

2.3.6 Linear Solvers

In the special case of solving mechanical systems, at the last step the problem is always reduced to a linear equation system with a system matrix M as seen above. Independent of the used formulation (FEM or particles), the matrix is usually sparse, symmetric, and positive definite. As a consequence, methods that take these properties into account to find the solution are of high importance. The aforementioned conjugate gradient method based on the Krylov subspaces is such a method which will be introduced here. Special attention will also be given to the convergence of the method with respect to preconditioning and the error bounds.

2.3.6.1 Krylov-Subspace Methods

The (orthogonal) Krylov subspace method is a projection method for finding the solution to a regular equation system of the form

$$A\mathbf{x} = \mathbf{b} \tag{2.151}$$

with a regular matrix $A \in \mathbb{R}^{n \times n}$ and $\mathbf{b} \in \mathbb{R}^n$. The Krylov Subspace K_m is defined as

$$K_m = K_m(A, \mathbf{r}_0) = span\{\mathbf{r}_0, A\mathbf{r}_0, \ldots, A^{m-1}\mathbf{r}_0\} \tag{2.152}$$

with

$$\mathbf{r}_0 = \mathbf{b} - A\mathbf{x}_0 \tag{2.153}$$

where $\mathbf{x}_0$ is the initial guess to the solution. The conjugate gradient method finds the optimal approximation $x_m \in x_0 + K_m$ to the solution $A^{-1}\mathbf{b}$ in terms of the orthogonality condition of the projection method

$$(b - A\mathbf{x}_m) \perp K_m \tag{2.154}$$

with any $\mathbf{x}_m \in \mathbf{x}_0 + K_m$ and $\mathbf{x}_0 \in \mathbb{R}^n$.

The CG-method requires A of (2.151) to be symmetric and positive definite. Let $\mathbf{v}_1, \ldots, \mathbf{v}_m \in R^n$ be the basis vectors of the Krylov Subspace K_m and $V_m = (\mathbf{v}_1, \ldots, \mathbf{v}_m) \in R^{n \times n}$. It follows that $V_m^T A V_m$ is regular and the resulting projection is given by

$$P_m = I - A V_m (V_m^T A V_m)^{-1} V_m^T \tag{2.155}$$

For the estimator of the error vector $\mathbf{e}_m = A^{-1}\mathbf{b} - \mathbf{x}_m$ and the residual vector $\mathbf{r}_m = A\mathbf{e}_m$ we have

$$\|\mathbf{e}_m\| \le \|A^{-1} P_m\| \min_{x \in P_m^1} \|p(A)\mathbf{r}_0\| \tag{2.156}$$

$$\|\mathbf{r}_m\| \le \|P_m\| \min_{x \in P_m^1} \|p(A)\mathbf{r}_0\| \tag{2.157}$$

where P_m^1 denotes the set of polynomials of maximal degree m for which $p(0) = I$ holds. In general, with $m = n$ follows the regularity of V_m such that $P_m = 0$ holds and the exact solution is found with (2.156) and (2.157). Thus, (orthogonal) Krylov subspace methods can be seen as direct scheme to solve (2.151). But practically, computers are of limited precision in their computations and introduce rounding errors whereby the solution is not necessarily found in n steps. As a result, these methods are used in an iterative manner, which is also of special interest in this work.

2.3.6.2 Conjugate Gradient-Method

Let $A \in \mathbb{R}^{n \times n}$ be symmetric and positive definite then it follows that for

$$F : \begin{cases} \mathbb{R}^n \to \mathbb{R} \\ \mathbf{x} \mapsto \frac{1}{2}\langle A\mathbf{x}, \mathbf{x} \rangle - \langle \mathbf{b}, \mathbf{x} \rangle \end{cases} \tag{2.158}$$

the solution to $A\mathbf{x} = \mathbf{b}$ is $\hat{\mathbf{x}} = \min_{x \in \mathbb{R}^n} F(\mathbf{x})$.

The minimum of F is found by subsequently searching for local minima in designated directions $\mathbf{p} \in \mathbb{R}^n$. These search directions are determined by the condition of being locally optimal w.r.t. to F, i.e. by choosing the negative gradient, and by enforcing orthogonality $A\mathbf{p} \perp K_m$ in each step. The latter ensures that the search directions are globally optimal w.r.t. to the space K_m. The negative gradient of the function F is

$$-\nabla F(\mathbf{x}) = -\frac{1}{2}(A + A^T)\mathbf{x} + \mathbf{b} = \mathbf{b} - A\mathbf{x} = \mathbf{r} \qquad (2.159)$$

The residual vectors $\mathbf{r}_0, \ldots, \mathbf{r}_m$ lead to the search directions

$$\mathbf{p}_0 = \mathbf{r}_0 \mathbf{p}_m = \mathbf{r}_m + \sum_{j=0}^{m-1} \alpha_j \mathbf{p}_j \qquad (2.160)$$

From the orthogonality condition follows

$$0 = \langle A\mathbf{p}_m, \mathbf{p}_i \rangle = \langle A\mathbf{r}_m, \mathbf{p}_i \rangle + \sum_{j=0}^{m-1} \alpha_j \langle A\mathbf{p}_j, \mathbf{p}_i \rangle \qquad (2.161)$$

and with $\langle A\mathbf{p}_j, \mathbf{p}_i \rangle = 0$ for $i, j \in \{0, \ldots, m-1\}$ and $i \neq j$ one obtains the α_i from

$$\alpha_i = -\frac{\langle A\mathbf{r}_m, \mathbf{p}_i \rangle}{\langle A\mathbf{p}_i, \mathbf{p}_i \rangle} \qquad (2.162)$$

For the computed search directions, one has to find the value λ, which minimises F along $\mathbf{p}$. Formally, the optimum is at

$$\lambda_{opt} := \min_{\lambda \in \mathbb{R}} F(\mathbf{x} + \lambda\mathbf{p}) = \frac{\langle \mathbf{r}, \mathbf{p} \rangle}{\langle A\mathbf{p}, \mathbf{p} \rangle} \qquad (2.163)$$

In a higher Krylov space K_{m+1} the solution can be incrementally improved by the formula

$$\mathbf{p}_m = \mathbf{r}_m + \frac{\langle \mathbf{r}_m, \mathbf{r}_m \rangle}{\langle \mathbf{r}_{m-1}, \mathbf{r}_{m-1} \rangle} \mathbf{p}_{m-1} \qquad (2.164)$$

$$\lambda_m = \frac{\langle \mathbf{r}_m, \mathbf{r}_m \rangle}{\langle A\mathbf{p}_m, \mathbf{p}_m \rangle} \qquad (2.165)$$

Finally, with the latter two equations (2.164) and (2.165) it is possible to start with a minimal Krylov space K_m and iteratively refine the solution by using a higher space until the residual vector is below an error bound. In summary, based on the formulas the algorithm then works as presented in Algorithm 2.1.

In the worst case the algorithm needs $n - 1$ steps to converge to the solution. But such a rough estimation is certainly useless in justifying the use of the CG method. As the projection strongly depends on the matrix A, it is important to find the maximal iteration steps w.r.t. the chosen error tolerance ε_{tol}. With the sequence

---

Algorithm 2.1 Iterative Conjugate Gradient Method

$\mathbf{r}_0 \leftarrow \mathbf{b} - A\mathbf{x}_0$ {initial guess $\mathbf{x}_0 \in \mathbb{R}^n$, e.g. $\mathbf{x}_0 \leftarrow \mathbf{0}$}
$\mathbf{p}_0 \leftarrow \mathbf{r}_0$
$\alpha_0 \leftarrow \langle \mathbf{r}_0, \mathbf{p}_0 \rangle$
$m \leftarrow 0$
while $(m \leq n - 1) \wedge (\alpha_m > \varepsilon_{tol})$ **do**
$\quad \mathbf{v}_m \leftarrow A\mathbf{p}_m$
$\quad \lambda_m \leftarrow \frac{\alpha_m}{\langle \mathbf{v}_m, \mathbf{p}_m \rangle}$
$\quad \mathbf{x}_{m+1} \leftarrow \mathbf{x}_m + \lambda_m \mathbf{p}_m$
$\quad \mathbf{r}_{m+1} \leftarrow \mathbf{r}_m - \lambda_m \mathbf{v}_m$
$\quad \alpha_{m+1} \leftarrow \langle \mathbf{r}_{m+1}, \mathbf{r}_{m+1} \rangle$
$\quad \mathbf{p}_{m+1} \leftarrow \mathbf{r}_{m+1} + \frac{\alpha_{m+1}}{\alpha_m} \mathbf{p}_m$
$\quad m \leftarrow m + 1$
end while

$\mathbf{x}_m$ of approximate solutions and A being symmetric and positive, definite, the error bound of $\mathbf{e}_m = A^{-1}\mathbf{b} - \mathbf{x}_m$ yields

$$\|\mathbf{e}_m\|_A \leq \left(\frac{\sqrt{\kappa(A)} - 1}{\sqrt{\kappa(A)} + 1} \right)^m \|\mathbf{e}_0\|_A \tag{2.166}$$

where $\kappa(A)$ denotes the condition number of matrix A respective to the $\| \cdot \|_2$ norm. We then get from (2.156) the upper bound for the maximal steps m to satisfy the error tolerance $\|\mathbf{e}_m\|_A \leq \varepsilon_{tol} \|\mathbf{e}_0\|_A$

$$m \leq \frac{1}{2} \sqrt{\kappa(A)} \ln\left(\frac{2}{\varepsilon} \right) + 1 \tag{2.167}$$

As one can see from the above equation and (2.156) and (2.157) that the condition number is of high importance for the speed of convergence. It is therefore desirable to have a matrix with a low condition number in order to have an efficient method for finding the solution to the problem. Unfortunately, one cannot influence the problem that formed the matrix to move into the desired direction. But we can find an equivalent matrix $\widetilde{A}$ with a reduced condition number. The transformation of a given equation system to an equivalent system with a reduced κ is called **preconditioning** and used in the CG method to decrease the number of iteration steps.

2.3.6.3 Preconditioning of the CG Method

It has been shown that the condition number is the main factor in the upper bound at a certain precision ε for the conjugate gradient steps. Hence, the obvious idea is to transform the problem $A\mathbf{x} = \mathbf{b}$ into an equivalent counterpart

$$P_l A P_r \mathbf{x}_p = P_l \mathbf{b} \quad \text{with } \mathbf{x} = P_r \mathbf{x}_p \tag{2.168}$$

Algorithm 2.2 Preconditioned CG-Method (PCG)

$\mathbf{r}_0 \leftarrow \mathbf{b} - A\mathbf{x}_0$ {initial guess $\mathbf{x}_0 \in \mathbb{R}^n$, e.g. $\mathbf{x}_0 \leftarrow \mathbf{0}$}

$\mathbf{z}_0 \leftarrow P\mathbf{r}_0, \quad \mathbf{p}_0 \leftarrow z_0$

$\alpha_0 \leftarrow \langle \mathbf{r}_0, \mathbf{p}_0 \rangle$

$m \leftarrow 0$

while $(m \leq n - 1) \wedge (\alpha_m > \varepsilon_{tol})$ **do**

    $\mathbf{v}_m \leftarrow A\mathbf{p}_m$

    $\lambda_m \leftarrow \frac{\alpha_m}{\langle \mathbf{v}_m, \mathbf{p}_m \rangle}$

    $\mathbf{x}_{m+1} \leftarrow \mathbf{x}_m + \lambda_m \mathbf{p}_m$

    $\mathbf{r}_{m+1} \leftarrow \mathbf{r}_m - \lambda_m \mathbf{v}_m$

    $\mathbf{z}_{m+1} \leftarrow P\mathbf{r}_{m+1}$

    $\alpha_{m+1} \leftarrow \langle \mathbf{r}_{m+1}, \mathbf{z}_{m+1} \rangle$

    $\mathbf{p}_{m+1} \leftarrow \mathbf{z}_{m+1} + \frac{\alpha_{m+1}}{\alpha_m} \mathbf{p}_m$

    $m \leftarrow m + 1$

end while

---

such that $P_l A P_r \approx I$ holds. Evidently, methods for solving linear equation systems are best suited to look for an approximate inverse of the matrix as preconditioner. But to consider the requirements of the CG-method, these method must not destroy neither the symmetry nor the positive, definiteness of the matrix. Otherwise the consistency and convergence of the CG-method would not be guaranteed. Preconditioners leaving the properties of the matrix untouched are therefore scalings, such as the incomplete Cholesky decomposition and symmetric splitting methods, e.g. Jacobi-(Relaxation), Richardson- and the symmetric Gauss-Seidel-(Relaxation-) method. The formal definition of such preconditioners can be found in [11]. The preconditioner can be inserted into Algorithm 2.2 to calculate the solution based on preconditioning.

2.4 Measuring Physical Properties

Finally, with the complete numerical simulation at hand, one has only to insert the correct parameters for any material into the calculations which brings the computed results into correspondence with reality. Therefore constitutive equations determine the amount of experiments needed to find an adequate model of the simulated material. Due to the fact that materials observed at small scale level are a conglomerate of atoms in a complex crystalline structure, materials can vary in their characteristics depending on several influencing factors forcing the structure to change. But fortunately, in most cases the working range of the used materials is very limited such that most properties remain constant. Especially, applications in mechanical engineering computing static configurations of civil structures, have determined important characteristics. Taking these properties into account (cf. Table 2.1), a sufficient description of structures and their reaction upon forces in finding the equilibrium can be given. In civil engineering the quantitative determination of elastic properties is possible with the measurement procedures illustrated in Fig. 2.9.

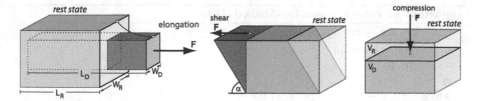

Fig. 2.9 Measurement of elastic, shear and bulk modulus

2.4.1 Textile Parameters

In addition to the previously illustrated material properties in the section before, textile materials have a more complex nonlinear behaviour. In other words, the working range in which the properties remain constant is very small. However, for our purpose of touching and stretching textiles it is sufficient to consider only nonlinear elasticity and omit other properties like creep or hysteresis. Contrary to materials like iron, concrete and wood is the regular structure of a fabric, it is composited by yarns which are themselves made of fibers. Fabrics are moreover distinguished in being woven or non-woven. Woven fabrics exhibit preferential directions due to their manufacturing process. More precisely, yarns are fixed in machine direction, the **warp** direction. A successive lifting and lowering of some of these yarns followed by a insertion of another yarn ("shooting"), **weft** direction, creates a weave pattern. Apparently, the weave structure has a strong influence on the mechanical characteristics of the fabric. Additionally, the yarns itself have different structures from the assembly of fibers, e.g. they can be twisted to a single yarn or two single yarns twisted together. Due to the complex manufacturing of fabrics, modelling textiles with linear elasticity would not reflect their real physical behaviour. Among the various parameters affecting the material behaviour those influencing the comfort of wearing are more important factors to the consumer. With the comfort being a very subjective assessment of the wearer, the textile industry has coined the term "subjective hand" or "fabric hand" as a general assessment of a set of subjective properties. A specially trained person has to assess the scale of each property in several manipulative tasks in order to characterise the fabric. As the assessment needs a skilled person to be present and is more importantly not objective, researchers correlate these subjective characteristics with the mechanical properties.

Tests for an objective assessment have been created resembling the subjective evaluation. With the Kawabata Evaluation System for Fabrics (KES-F) [9] named after its inventor Sueo Kawabata, such objective evaluation was made possible and has become the de facto standard in fabric assessment.

He constructed machines dedicated to run predefined test procedures yielding quantities related to the fabric properties. Thus the subjective properties are transformed into objective quantifiable parameters with a mechanical correspondence. As the subjective factors are influenced by the mechanical behaviour of fabric, the measurements taken by the machines cover all important mechanical properties,

Table 2.2 Characteristic values obtained by the KES-F system [9]

Property	Symbol	Characteristic	Unit
Tensility	LT	Linearity	gf[a] cm/cm^2
	WT	Tensile energy	
	RT	Resilience	%
Bending	B	Bending rigidity	gf cm^2/cm
	2HB	Hysteresis	gf cm^2/cm
Shearing	G	Shear stiffness	gf/cm degree
	2HG	Hysteresis at 0.5°	gf/cm
	2HG5	Hysteresis at 5°	
Compression	LC	Linearity	gf cm/cm^2
	RC	Resilience	%
Surface	MIU	Coefficient of friction	
	MMD	Mean deviation of MIU	
	SMD	Geometrical roughness	µm
Weight	W	Weight per unit area	mg/cm^2
Thickness	T	Thickness at 0.5 gf/cm^2	mm

[a]gf: gram force = 0.00980665 N

e.g. bending, tensility, shear. Table 2.2 shows the various parameters evaluated by the machines of the KES-F.

The complete extraction of the fabric's properties is done in four units. The first unit as shown in Fig. 2.10(a) measures the elastic behaviour of the fabric. A specimen 20 cm wide and 5 cm high is placed horizontally into an attachment which fixes the fabric at top and bottom of the machine. The attachment separates top and bottom ending with a constant speed yielding a one-directional strain at constant rate. The resulting force is plotted with respect to the strain. The strain increases until 500 gf/cm^2 in resistance force has been reached. At this limit point the movement is reversed and the recovery of the specimen is measured.

The identical setup is used in the measurement of shearing. Here, the specimen is sheared by a parallel movement of one attachment relative to the other. During the movement the resistance force is plotted until a shearing of 8° is reached. Analogous to the tensility measurement, by reversing the movement the recovery of the fabric is recorded as well. Reaching of the initial position, the machine continues to move in the opposite direction and measures the shearing behaviour in negative angle.

With the second unit as depicted in Fig. 2.11(a), the fabric's bend behaviour is evaluated. A specimen of 20 cm height and 1 cm width is hanging vertically in the machine's attachment, fixing left and right ends. In this test, the machine creates a curvature over the 1 cm width by moving one attachment on a circular path around the other starting from 0 to 90 degrees. The torques produced by bending are recorded by a sensor inside the rigid attachment.

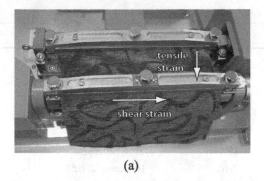

(a)

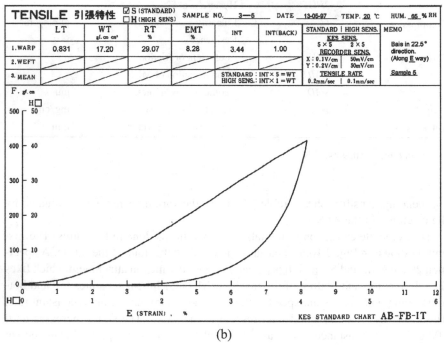

(b)

Fig. 2.10 (a) Specimen in shear and tensile strain. (b) Plot of tensile forces

Compression and thickness of a specimen is measured in a third unit. Two circular steel plates are pressed together to find the compressional resiliency of the material. In the evaluation process the position is plotted with respect to the pressure. The process reverses when 50 gf/cm^2 have been reached. Thickness of the material is simultaneously measured when the pressure attains 0.5 gf/cm^2.

The last block is used to determine the surface features of the fabric, the geometrical roughness and the friction. As the geometrical roughness is defined as the variation of the height of the surface, the machine consequently records these variations by a probe moving over the specimen with a constant speed of 1 mm/s. The probe itself consists of a U-shaped steel wire with a diameter of 0.5 mm which is pressed

(a)

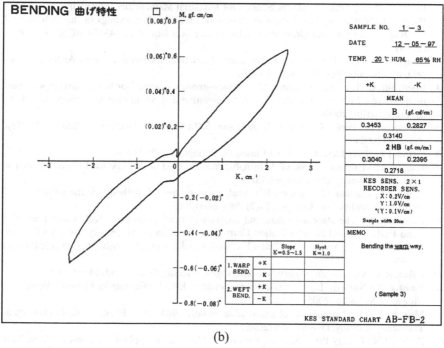

(b)

Fig. 2.11 (a) Defined bending of textile. (b) Moments vs. curvature graph

on the specimen surface with a spring force of 10 gf and stiffness of 25 gf/mm. To prevent strong indentation of the probe, the specimen is kept in tension by a force of 20 gf/cm. The plotting apparatus records the thickness variation during the movement for 20 s. At the same time, the friction is recorded as well. The friction probe is constructed similar to the surface probe, it only differs in the width, by consisting of more wires. The friction is measured as variation of the tangential force produced by the movement and the normal force of 50 gf. An additional filtering is used to compute the mean values in both variations (cf. [9] for details).

With the measurements made by the KES-F, it is possible to extract the strain-stress relationship of the material. But still, simplifications have to be introduced to consider the measurements to be complete. Thus, we assume an orthotropic material behaviour of woven textiles, restricting ourselves to fabrics with symmetric weave patters, independence of the deformation modes (tensile and shearing), where the nonlinear response is determined by the stress-strain curves obtained in each single weft/warp and shear direction (as seen in Figs. 2.10(b) and 2.11(b)).

References

1. Ascher, U.M., Petzold, L.R.: Computer Methods for Ordinary Differential Equations and Differential-Algebraic Equations. Society for Industrial Mathematics, Philadelphia (1998)
2. Baraff, D., Witkin, A.: Large steps in cloth simulation. In: Proceedings of the 25th Annual Conference on Computer Graphics and Interactive Techniques, pp. 43–54 ACM, New York (1998)
3. Bonet, J., Wood, R.D.: Nonlinear Continuum Mechanics for Finite Element Analysis. Cambridge University Press, Cambridge (1997)
4. Eberhardt, B., Etzmuß, O., Hauth, M.: Implicit-explicit schemes for fast animation with particle systems. In: Eurographics Computer Animation and Simulation Workshop, vol. 2000, Springer, Berlin (2000)
5. Goldstein, H., Poole, C., Safko, J., Addison, S.R.: Classical Mechanics. Addison-Wesley, Reading (2002)
6. Gould, P.L.: Introduction to Linear Elasticity. Springer, Berlin (1994)
7. Hauth, M.: Visual simulation of deformable models. PhD thesis, Eberhard-Karls-Universität Tübingen, Germany, Dissertation (2004)
8. Hauth, M., Etzmuß, O., Straßer, W.: Analysis of numerical methods for the simulation of deformable models. Vis. Comput. **19**(7), 581–600 (2003)
9. Kawabata, S.: The standardization and analysis of hand evaluation. Technical report, The Hand Evaluation and Standardization Committee, The Textile Machinery Society of Japan. Osaka Science and Technology Center Bld., 8-4, Utsubo-1-chome, Nishi-ku, Osaka 550 Japan (1980)
10. Lakes, R.S.: Viscoelastic Materials. Cambridge University Press, Cambridge (2009)
11. Meister, A.: Numerik Linearer Gleichungssysteme: Eine Einführung in Moderne Verfahren. Vieweg, Wiesbaden (2005)
12. Mezger, J.: Visual simulation of deformable models. PhD thesis, Eberhard-Karls-Universität Tübingen, Germany, Dissertation (2008)
13. Reddy, J.N.: Energy Principles and Variational Methods in Applied Mechanics. Wiley, New York (2002)
14. Reddy, J.: An Introduction to the Finite Element Method. McGraw-Hill, New York (2006)
15. Volino, P., Magnenat-Thalmann, N.: Accurate garment prototyping and simulation. Comput. Aided Des. Appl. **2**(1–4) (2005)
16. Volino, P., Magnenat-Thalmann, N.: Implicit midpoint integration and adaptive damping for efficient cloth simulation. Comput. Animat. Virtual Worlds **16** (2005)
17. Volino, P., Magnenat-Thalmann, N.: Accurate anisotropic bending stiffness on particle grids. In: International Conference on Cyberworlds, CW'07, pp. 300–307 (2007)

Chapter 3
Haptic Interaction

Virtual Reality (VR) systems have evolved in the way that the visual immersion provided by technologies like CAVEs [10] or head mounted devices is nearly perfect. But since the visual quality reached a satisfactory level, a deficiency became apparent in the interaction with the virtual environment. The absence of touch information during manipulative task made interactions difficult and cumbersome. Moreover, the estimation of material characteristics where necessary, e.g. medical surgery simulation, is nearly impossible (cf. [31]). Therefore, VR systems requiring the user to make complex interactions are augmented by a touch feedback. In the following the issues and concepts in providing a touch feedback which is also known as haptic interaction will be explained.

3.1 Preliminaries

The interaction is usually conceived by the immersion of the user in a virtual environment (VE) with the direct control of an object, either a virtual tool or a virtual part of himself. Then the VR system tries to compute the forces or positions that a user would receive due to his actions as an immediate result of the current configuration in the VE. The computed signals are *displayed* by a so-called **haptic device**. For the computation as well as for the design of such devices, it is useful to consider the characteristics of our touch sense to provide the illusion of touch is provided by a haptically augmented VR system.

3.1.1 Touch Perception

The human perception consists of five senses, one of them being the sense of touch. In contrast to the other senses, the touch sense is not perceived by a centralised organ. Instead, it consists of sensors named receptors distributed through our entire body encoding perceptional information upon stimuli. The final perception in our

G. Böttcher, *Haptic Interaction with Deformable Objects*,   51
Springer Series on Touch and Haptic Systems,
DOI 10.1007/978-0-85729-935-2_3, © Springer-Verlag London Limited 2011

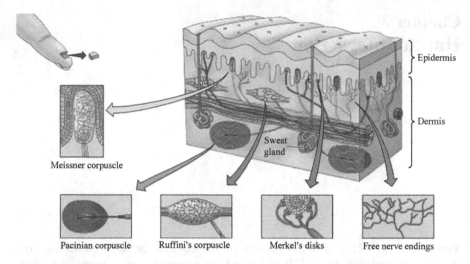

Fig. 3.1 Mechanoreceptors involved in the skin touch perception (image from Purves et al. [30])

brain is generated by this sensory network, the network and the signal processing called somatensory system. The system is also able to produce information to further sensory modalities, e.g. temperature, pain and body posture. The receptors are distinguished by their location, namely cutaneous and inner body. The former are sensitive to mechanical input due to skin deformation and vibrations caused by tangential movement. In total there are four different mechanical receptor types responsible to create the skin touch sensation (shown in Fig. 3.1). The inner body receptors regarding the touch sense are located in our muscles, tendons and joints providing feedback on posture and forces. The latter receptors are of special importance in the following discussion.

In the terminology of perception, the sense of touch is also referred to as the **haptic** sense. On the basis of the aforementioned locality and the receptor functions in the haptic perception, a discrimination is made by the terms **tactile** and the **kinesthetic** perception addressing the specific receptors. Consequently, the tactile perception considers only small scale forces by slight touch or surface movement and allows us to feel the smoothness or bumpiness of structures. The kinesthetic perception is responsible for perceiving gross mechanical forces like the weight, resistance of objects or simply the position of our extremities. Although the haptic perception or haptics respectively are considered to be equal to tactile perception, the term is usually extended to cover also the kinesthetic perception. The methods and analysis in the work accounts mainly for the kinesthetic sense.

Experiments on the touch perception in terms of interface design was conducted by Tan et al. [32] in order to find the bandwidth and thresholds. They found out that vibrations during the control of a virtual object in contact or free motions can significantly degrade the touch illusion. Our somatensory system is able to perceive vibrotactile stimuli up to 1000 Hz, but the upper limit in force control is 20 to 30 Hz. Hayward et al. [17] conclude further that the human force perception ranges from 1 mN to at least 10 N.

Fig. 3.2 Commercial device
used for gaming

3.1.2 Haptic Devices

Haptic devices are in their design very similar to robot arms. Both are typically
constructed by motors actuating the joints of the arm to control either position or
force of the end of kinematic chain, termed **end effector**. The main difference lies
solely in their use. The robot arm is designed to manage manipulative tasks with a
specific end effector, e.g. a gripper. A haptic device is, however, designed to transfer
forces to the user by its end effector, e.g. with a thimble.

For this reason, haptic devices emerged from the field of robotics. The early
application was to create an interface for remote control of robot arms with sensory
feedback. The devices were designed to give the user the ability to perceive forces
from his interaction with a remote location, i.e. the user "feels" the resistances to
actions of the robot. Research problems of such applications were mainly related
to the processing the sensor data and ensuring stability induced by the transmission
delay, such that the user is not harmed by the device. The next major step was made
by replacing the sensoric information obtained from a remote site with artificial data
generated by a computer. This was the beginning of haptic interaction.

Together with the commercial availability of such devices at moderate prices like
the Novint Falcon® shown in Fig. 3.2 and the research in more sophisticated virtual
realities, the devices gained attraction to the computer graphics community. Since
then the applications of such devices has been expanded to the use in virtual reality
environments as an interactive tool for training, simulation, and exploration.

One of the main properties of a haptic device is the degree of freedom (DOF) in
which it can provide a force, e.g. force-feedback joystick is 2DOF-device whereas
a stylus operated device exhibits 3DOF as shown in Fig. 3.3. Another property is
the number of contact points at which the forces are displayed to the users. In the
latter example, it would be either the entire hand or the two fingers holding the
stylus. For the purpose of displaying pinch-grasp interaction forces, devices should
have 6DOFs (force and torques) and two contact points. Devices differ further in
their control by being either **admittance** or **impedance** controlled as proposed by
Yoshikawa et al. [33]. In the former, the user applies a force to the device and the
device moves w.r.t. to the VE. The latter is controlled by the movement of the user
and outputs a force commanded by a VE. Both principles feature high gains in

Fig. 3.3 Stylus operated
device for 3d modelling

the motor control in switching from free motion to contacting (impedance) or the opposite way around (admittance). These gains pose a problem in stability of the control of the device resulting in vibrations or high forces. This and other problems in stability will be addressed in the next sections. In the following discussion only the class of impedance controlled devices which are more widely spread because of their lower price and easier control will be considered.

Another issue that all devices have in common is the need of concealing their presence in free movement during operating. An ideal device has to be completely imperceptible to the user and it should not create forces in static and dynamic cases. But since the device consists of mechanical parts with a certain weight, the device possesses inertia of mass and friction in the joints, which can only be inhibited to a certain degree by active control. These undesired forces are additionally felt by the user. Therefore, physical properties of the device play an important role in the quality of resulting force feedback. Hayward et al. [18] gives a detailed overview on the issues in device construction.

The design of a devices is not only motivated by the aforementioned properties but also by the force transmission. The higher the physical force generated by device is, the more power is needed, which leads to bigger motors and thus heavier devices. Additionally, with the force applied to the user, a counteracting force is inherently given by the physical laws needs to be dissipated by a grounding. Therefore, different concepts of force transmission were invented having benefits in their special use, e.g. user mounted (e.g. hand-exoskeleton as from Bergamasco et al. [4]) are best suited for mobile applications requiring only forces on fingers while maintaining freedom in motion, or grounded devices (e.g. desk-mounted from Massie et al. [26]) allowing to create high forces and reducing the fatigue of user having no weight to carry. For the purpose of the presented VR system, the desk-mounted devices are best suited since no mobility is needed but high forces (up to 20 N and more).

As previously sketched in the introduction to haptic rendering, displaying forces is different from the visual display from the technical point of view. from. Here, the difference manifests in the direct feedback to the transmitted force signal. Due to this circumstance, the user can be seen as a component in a signal loop consisting of the device and the rendering software providing the VE, called **control loop**. In such a loop, a device is stated to be **passive** if it does not amplify the signal coming

Fig. 3.4 Block diagram of different controls

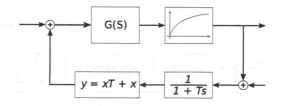

from the user or the VE. This is a necessary condition for the stability of the control loop. It can however not be inherently assured by the device and depends on the VE. Moreover, a device is limited by its mechanical and electrical properties such that it cannot be considered nor constructed as an ideal transmitter. This implies a non-linear relation between commanded force and displayed force.

To conclude, one has to understand the limit of the output of the device and how the signal is affected by such limits. In the following section, some common knowledge will be given needed to constitute an analysis method for finding such the limits will be given.

3.1.3 Control Theory

The principles of control theory deal with dynamical systems and their influence on each other in control loops, whereas a dynamical system is processing and transmitting data. An input signal is treated as the cause, which results in a temporal successive effect being the output signal. Thus, dynamical systems create a relation between cause and effect, respectively input and output signal. The control theory is considered to describe the control loop mathematically. It provides tools to develop controls having a specific behaviour approximating the desired behaviour. For further reading on the general topic of feedback and control the book of Stefano [11] is recommended.

The analysis of a dynamical system usually begins with a decomposition of the system into smaller connected systems with a simple, known behaviour. This allows to follow the signal and its transformation within the system. Such a decomposition is represented by a block diagram (e.g. Fig. 3.4) showing the flow of data and the operation of each block. The function of each block is indicated by a function graph, a mathematical description or special function. The arrows showing the direction of information flow can also specify the type of output and the relation by having a description. The diagram visualises the signal transform from input to output of the control loop.

In general, a control (loop) adjusts an output y based on the input w to a desired value y_s (setpoint) where output is called controlled variable. Two types of controls exist, the feedback and feed forward control. Within feedback control the controlled variable is constantly measured in order to eliminate the discrepancy to the setpoint. Here, the output y is fed back to the controller. Due to the direct link between control variable and input, the control creates **closed loop**.

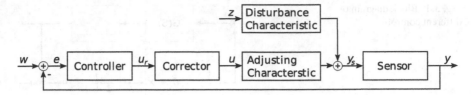

Fig. 3.5 Basic structure of a control loop

A feed forward control adapts the setpoint to a changed situation before a discrepancy is measured. These changes are modelled by disturbances with a predictive model for the signal variation of the setpoint. The drawback of a feed forward control is that the influence of disturbances are mostly not a priori known. Feed forward controls do usually not directly link the controlled variable with the input. Hence, this type of control creates an **open loop**. Figure 3.5 shows the principle block diagram of a closed loop control.

In the description of dynamical systems different mathematical models were developed. These models are suitable for different kind of systems which are separable into the categories *linear* and *non-linear*, *time-continuous* and *non-continuous* or *time-invariant* and *non-time-invariant*. The behaviour of a system is defined by equations describing the static or dynamic case. Abruptly changing input variables will normally not result in an immediate change of the output. Instead, it will rather adapt slowly to the new input. The stable output value reached after the adaptation is described by the static behaviour. The transient change is given by the dynamic behaviour. This behaviour is expressed mathematically by an operator T, creating a relation between input $x(t)$ and output $y(t)$ value.

$$y(t) = T[x(t)] \tag{3.1}$$

If the corresponding operator T follows the principle of superposition, then the system is determined as being **linear**

$$\sum_{i=1}^{n} k_i y_i(t) = T\left[\sum_{i=1}^{n} k_i x_i(t)\right] \tag{3.2}$$

where the coefficients k_i are the proportionality factors.

A system is called **time-invariant**, if the system parameters are constant over time and the system reacts equally for the same inputs at any time. Formally, the system suffices the mathematical condition

$$y(t - t_0) = T[x(t - t_0)] \tag{3.3}$$

which implies that a time shift of the input signal results in the same shift of the output with equal response. Systems processing signals at any time are called **continuous**, whereas systems controlling at certain time steps are called **discrete**. Although in reality many quantities are continuous, e.g. position or velocity, some controls are realised by a computer having only the capability of time and discretely processing value. Hence, it is unavoidable to convert continuous signals into discrete ones and vice versa. Those conversions are shown later in the actual context of haptic rendering.

A system is causal, if the effect occurs after the cause. The latter class of systems corresponds to all systems in reality. The input signal begins as common agreement at $t = 0$, which is formally equivalent to

$$x(t) = 0, \quad \forall t < 0$$

For the causal system it also implies

$$y(t) = 0, \quad \forall t < 0$$

Systems of linear and continuous behaviour can be mathematically expressed by differential equations. Additionally, they can be described by their **step response** $h(t)$. It is defined as the output signal of unit step input signal by

$$\sigma(t) = \begin{cases} 1, & \text{for } t \geq 0 \\ 0, & \text{otherwise} \end{cases}$$

The graph of the step response is a common description within a block diagram. Moreover, the system can be defined by their weighting function or **impulse response** $g(t)$. It corresponds with the output signal resulting from a Dirac impulse $\delta(t)$ at the input. Since

$$\delta(t) = \frac{d}{dt}\sigma(t), \tag{3.4}$$

it follows that

$$g(t) = \frac{d}{dt}h(t) \tag{3.5}$$

Thus the output value of a system can be described by the formula below if the impulse response is known, see also [11].

$$y(t) = \int_0^t g(t - \tau)x(t)\,d\tau \tag{3.6}$$

For the analysis of systems governed by linear differential equations with constant coefficients, the **Laplace transform** is used. It transforms the function $f(t)$ into the frequency domain with the following rule

$$F(s) = \int_{\infty}^{\infty} f(t)e^{-st}\,dt, \quad s \in \{+i\omega\} \subset \mathbb{C} \tag{3.7}$$

In case of having a causal system the integral has to converge. The convergence is only guaranteed for the real part of the complex plane $Re(s) > \sigma_0$ where σ_0 denotes the abscissa of convergence. Another property of the Laplace transform is the transition of convolutions in the time domain to multiplication in the frequency domain. This allows to compute the Laplace transform of the output $Y(s)$ simply by the multiplication of the transform of the impulse response $G(s)$, also known as **transfer function**, and the input signal $X(s)$.

$$Y(s) = G(s)X(s) \tag{3.8}$$

The transfer function $G(s)$ is a rational function which can always be represented in the following factorised form

$$G(s) = \frac{Y(s)}{X(s)} = \frac{b_0 + b_1 s + \cdots + b_m s^m}{a_0 + a_1 s + \cdots + a_m s^m} = k_0 \frac{(s - s_{N_1})(s - s_{N_2}) \cdots (s - s_{N_m})}{(s - s_{P_1})(s - s_{P_2}) \cdots (s - s_{P_n})} \tag{3.9}$$

where the roots s_i^R and the poles s_i^P are only of real or complex conjugated value since the constants a_i and b_i are because of physical reasons real valued. Thus a linear, time-invariant system without dead time can be fully described by its roots and poles in the complex plane and gain factor k_0.

A system is called **stable**, if a bounded input signal results in a bounded output signal. Formally, this is true if the following holds

$$\lim_{t \to \infty} g(t) = 0 \tag{3.10}$$

Is the limit further real valued, the system is called **critically stable**. Otherwise, if no limit exists, the system is **instable**. In the frequency domain, the property of having a stable system is satisfied if all poles have a negative real part. In case of simple poles having a vanishing real part and all other poles are negative, the system is critically stable. For having an instable system, one pole has a positive real part or the real part of a double pole vanishes.

Although it is desirable to analyse haptic devices in terms of stability with the tools of control theory, it is not so often made as one should think. The main problem for this lies in the coupling with the user. Since user is integrated into the control loop and his transfer function is unknown, a prediction of the behaviour of the device is theoretically impossible. The user's behaviour changes in time and is also different from user to user. Therefore, the user is a system of non-linear and time variant behaviour and without experimental tests, the analysis of the haptic device's stability cannot be complete.

3.1.4 Signal Processing

For those experimental test, it is necessary to digitally record the output and inputs in order to assess the behaviour of the system. Obviously, it is always a discrete systems which has to be analysed as the sampling of the signals create discrete series of values. As before, discrete systems are discussed in the frequency domain because most conclusions can be drawn here. But the behaviour of the complete system considering the user as an element in the control loop is unknown. Problems in the discrete case arise when it has to be represented in frequency domain. Fortunately, these problems are of general nature and are dealt with the signal processing theory which is outlined in the following (see [29] for a comprehensive study).

At the beginning stands the transformation into frequency domain of the function $f(t)$. With the unitary scalar product in the function (vector) space $C(\mathbb{R}, \mathbb{C})$

$$\langle f, g \rangle := \int_{\infty}^{\infty} f(t)\overline{g(t)}\, dt \tag{3.11}$$

the function can be transformed into a new basis. The new basis to be used is defined by

$$b_\omega(t) := e^{i\omega t} \tag{3.12}$$

In this particular case of choosing the basis b_ω, the transformation is called **Fourier transform** and is, as one can see at (3.7), strongly related to the Laplace transform. In the study of the function in this basis is commonly denoted by $\mathcal{F}\{f\}$ and the strength of the signal at frequency ω is indicated by $\mathcal{F}\{f\}(\omega)$.

To adapt this scheme to discrete measurements two changes have to be made. First, one has to consider the limit in maximum frequency content in our discrete input signal. It can be clearly seen that there has to be a maximum as the signal has only finite input samples N in time T. The limit is called **Nyquist frequency** and is equal to the half of the sampling frequency $\frac{N}{T}$. This makes the calculation of frequencies above the limit dispensable. The second change limits the input to periodic sequences. A sequence of N values of samples in time domain corresponds to N values in the frequency domain.

By replacing the integration with a summation, the new scalar product

$$\langle f, g \rangle := \sum_{k=0}^{N-1} f(k)\overline{g(k)} \tag{3.13}$$

and the new basis

$$b_\omega(k) := e^{i\omega \frac{k}{N}} \quad \text{with } \omega = 2\pi m, \ m = 1 - N/2 \tag{3.14}$$

reflect those changes. Since the new formula considers only discrete frequency components of the signal, the transformation method is called **Discrete Fourier Transform** (DFT). It is widely used in an optimised fashion by exploiting the factorability of N named **Fast Fourier Transform** (FFT). Hereby, the complexity of the algorithm reduces from $\mathcal{O}(N^2)$ to $\mathcal{O}(N \log N)$.

A problem arises by the new method as it is restricted to periodic sequences. Signals without any periodicity may result in frequency components that were not necessarily present in the original input signal. These artefacts are mostly caused by discontinuities at the border of the section. To remove such artefacts a so-called

Fig. 3.6 Example of a
window function (hamming
window)

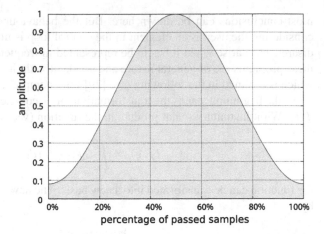

window function is applied prior to the transformation. The window function is
multiplied with the sampled sequences and reduces the signal near the borders by
becoming zero. The window function inherently describes the sensitivity of the basis
elements to nearby frequencies and thus cannot be arbitrarily defined. Transforming
the signal without using a window function is equivalent to a rectangular window,
which has a wide spectrum sensitivity.

A concept similar to the window functions are **filters**. The difference here lies in
the application of a filter function, it is contrary to the window function multiplied in
the frequency domain. The spectrum of the filter is also called **frequency response**
and its inverse transform the **impulse response** being the output of an ideal pulse.

For making use of filters within the time domain, one has to perform a convo-
lution of the sequence with the impulse response of the filter. The use for filters in
the time domain is needed in efficient hardware implementations with small delays
having only the sampled sequence in a buffer. The calculation treats each sample as
a pulse and their responses are superpositioned with responses of the previous samples in the output. Since the filters needs to be causal, the impulse response affects
only the output in the future.

In analogy to continuous systems of the previous chapter, discrete systems can
be transformed into frequency domain by the **z-Transform** of their finite difference equations. The transformation rule is comparable to the Fourier transform for
sample sequences. With the change of the basis to powers of $z - 1$, denoting the
delay by one sample, the sequences become in the transform polynomials. The filter
or respectively the transfer function is hereby the ratio of output and input signal
expressed in a rational function.

The aforementioned kind of filters working in the discrete time domain are called
finite impulse response (FIR) filter, as they have a finite response to an occurring
impulse. The corresponding counterpart is the **infinite impulse filter** (IIR). These
filters have an internal feedback creating an infinite response to the input. The feedback in the calculation in the time domain is realised by adding a weighted sum of
the previous output values to the current value. Since both filter types are described

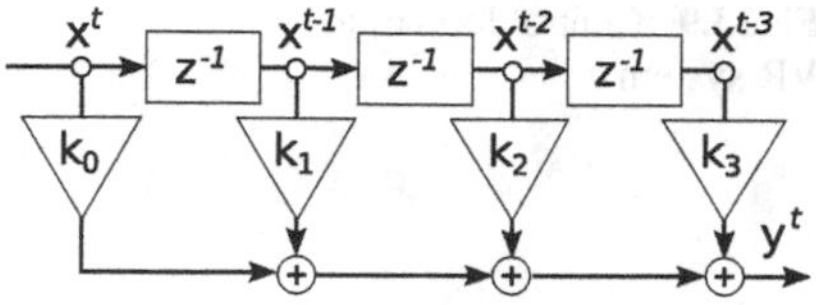

Fig. 3.7 Simple (moving average) FIR filter

Fig. 3.8 IIR filter by weighted feedback

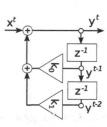

by difference equations, they can be seen as a discrete systems. Figures 3.7 and 3.8 illustrates the function of these filter types in a block diagram.

The region of stability in the complex plane is the unit circle and all poles must therefore lie in this circle. If a pole is on or outside the circle, the filter is instable. Due to their multiplicative operation in frequency domain, filters can be factorised.

If a Laplace transformed function of the filter is known an approximation of the z-transformed function can be given by using the Bi-linear Transform. Thus the s-plane has been mapped to the z-Plane, which can be achieved by

$$s = \frac{2}{T} \ln(z) \tag{3.15}$$

One obtains a substitute for s by representing the natural logarithm function with its first-order approximation:

$$s = \frac{2}{T} \ln(z) = \frac{2}{T} \left[\frac{z-1}{z+1} + \left(\frac{z-1}{z+1}\right)^3 + \left(\frac{z-1}{z+2}\right)^3 + \cdots \right] \approx \frac{2}{T} \frac{z-1}{z+2} \tag{3.16}$$

3.1.5 Stability

To analyse the system in the context of control theory, modelling of the components is needed. In [9, 14] and [24] are slightly different approaches presented having the same goal in common.

The user forms a system with an unknown transfer function $U(s)$. Despite the unknown function, the analysis is made possible by choosing a set of transfer functions with different characteristics. A force-feedback device is then assumed to be an idealistic point mass governed by its motion equation (2.2). The force $\mathbf{F}(t)$ is treated as input signal and the position $\mathbf{x}(t)$ is defined as the output. With the system considered as spring-mass-oscillator, the motion is described by a second-order ordinary differential equation. Then, the transfer function, the response to the forced

Fig. 3.9 Control loop of a
VR system

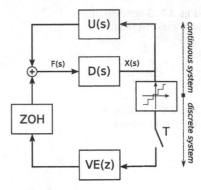

Fig. 3.10 Energy leaks at a
virtual wall

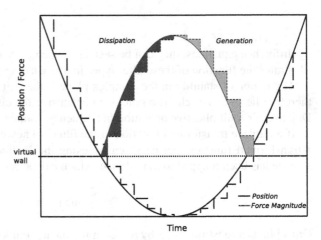

oscillations of the system, is thus defined by a Laplace transform and leads to the
theoretical function

$$F(s) = ms^2 X(s) + bs X(s) + k X(s)$$

$$D(s) = \frac{X(s)}{F(s)} = \frac{1}{ms^2 + bs + k} \tag{3.17}$$

By looking closer at the used devices, the idealistic point of view may shift a little
to reflect the mechanical design of the device. Figure 3.9 shows the control loop of
the entire system. The user and the device are continuous whereas the virtual environment is working with discrete time steps and values. Thus its transfer function
using the position as input is given in z-transform by $VE(z)$. The input position
is discretised by the sampling and the output force is made continuous by a hold
element of zero-order (ZOH).

In the analysis of a control loop stability, very clear and simple stress test has been
defined in [14] by simulating the contact with a stiff virtual wall. As virtual wall is of
high stiffness, the system has to react with very high impedances which may result
instable or vibrational situations due to energy leaks as illustrated in Fig. 3.10. The

first leak is the time discrete simulation where the force is proportional to the penetration of the wall. This strongly monotonic increasing function is transmitted to the device and is hold until the next update. But since the user is a continuous system, it does not hinder him to move the device further into the virtual wall. As the force is being kept constant in this period, the energy which would have normally been produced by the motion is higher than actually. Seen from the users perspective, his energy is dissipated by the discrete system. The opposite effect happens when the user moves out of the wall. In this case the force stays at a high value although the penetration depth of the user is much lower. Thus, the device adds more energy to the complete system than it is actually needed for the rendering.

Another reason for energy leaks is the inability of the system in knowing the exact ingression or egression time of the user in the virtual wall. In both situations the device reacts to slow such that the energy is dissipated at the ingression and created at the egression.

By increasing the sampling frequency of the system it is possible to reduce the aforementioned effects. But due to limited computing power, such an approach is certainly undesirable and thus not suitable for complex simulations. Instead of that [14] proposes two solutions to the problem. In the first approach, the position is extrapolated for a the half of a time step in advance in order to compute a mean in force between the actual and the next time step.

The second approach suggest a design of a digital control that compensates the effect of the hold element keeping the time discrete value constant over one time step. But since the transfer function of the user is unknown, the approach is not straight forward and needs a variety of experimental tests to estimate the transfer functions. As the VR system presented in this work has to be flexible, the latter approach is unsuitable.

3.1.6 Virtual Coupling

General stability of VR systems is only ensured if the system considers the behaviour of the used device. Since the device is part of the whole control loop of the system, its properties influence the haptic performance of the overall system. Therefore, the development of a haptic virtual reality system is mostly limited to a specific device.

To eliminate this restriction of having only one device or making the system independent of the used device, it is desirable to add another layer of abstraction between the virtual environment and the device. Instead of adapting the virtual environment to the device, the layer component is adjusted to the device. One of the main tasks of this component is also to satisfy the stability of the overall system by filtering those frequencies from the force signal that would render the system into an instable state. With the separation of the device from the virtual environment, a direct control of the user by the force rendering is not possible anymore. The haptic perception may on one side now differ from the simulated physic, but on the other it is ensured that

Fig. 3.11 Device as two-port
interface between user and
virtual environment

the rendering is always stable. Since the stability is the most important factor for a
authentic haptic feedback, it should therefore be favoured.

In [1, 2] and [9] the control loop is modelled by a two-port network having two
in- and outputs. The port condition to be satisfied by a two-port network is that the
inflow has to be of same magnitude as the outflow. By the modelling introduced in
the referenced articles, an analogy between electric circuits and mechanical structures is created. The voltage corresponds to the force and the current to the velocity.
As a common approach of electrical engineering, a one-port establishes a relation
between voltage and current with its transfer function. The user and the virtual environment are considered being a one-port which are connected by the haptic device
treated as two-port. The two-port links the one-ports with its transfer functions by
attaching the inputs of the one ports to the velocities and the outputs to the forces as
illustrated in Fig. 3.11

Mathematically, the relations are expressed by a 2×2 matrix along with the transfer functions in the frequency domain. In the ideal model of the two-port network
creating the link between velocity and force within the haptic device, the device is
described by

$$m_d \dot{v}_d + b_d v_d = F_h - F_d \quad \text{with } v_d = v_h \qquad (3.18)$$

and the relation

$$\begin{pmatrix} F_h \\ -v_d \end{pmatrix} = \begin{pmatrix} m_d s + b_d & 1 \\ -1 & 0 \end{pmatrix} \begin{pmatrix} v_h \\ F_d \end{pmatrix} \qquad (3.19)$$

To analyse the system in discrete time of the virtual environment, the continuous
transfer functions need to be discretised. By using the Bi-linear Transform one obtains an approximated z-Transform of these functions. This yields to

$$Z_d(z) = (m_d s + b_d)|_{s \to \frac{2}{T} \frac{z-1}{z+1}} \qquad (3.20)$$

And the transfer function of the assumed hold element (ZOH) is

$$ZOH(z) = \frac{1}{2}(1 + z^{-1}) \qquad (3.21)$$

Inserting the new functions into the device matrix leads to the equation

$$\begin{pmatrix} F_h \\ -v_d \end{pmatrix} = \begin{pmatrix} Z_d(z) & ZOH(z) \\ -1 & 0 \end{pmatrix} \begin{pmatrix} v_h \\ F_d \end{pmatrix} \qquad (3.22)$$

whereas the signals v_h and F_h remain time continuous. The stability criteria in [1]
for linear time-invariant two-port-networks allows to find a stable solution for the

Fig. 3.12 Mechanical
equivalence of virtual
coupling

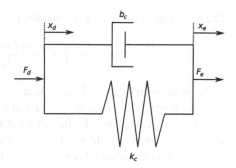

Fig. 3.13 Virtual coupling of
device in haptic control loop

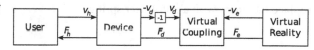

coupling by fulfilling

$$\cos(\varphi(ZOH(z))) \geq 1 \tag{3.23}$$

with $\varphi(z)$ as the phase shift of the system. But this holds only for a phase shift
of zero. As the hold element induces a phase shift, the condition cannot be met.
Although the condition cannot be satisfied, it does not mean the system cannot be
stable. The right conclusion is that the system cannot be stable under all circum-
stances.

In order to improve stability, a two-port element is added to the circuit (Fig. 3.13)
representing the virtual coupling. In the aforementioned article of Adams et. al, it is
suggested to choose the transfer function for the coupling such that it is a mechanical
equivalent of a spring-damper element being in series with the device (as illustrated
in Fig. 3.12). Thus the forces from the virtual environment are directly transmitted
whereas on the other hand the position is changed. The constants of the spring k_c
and the damper b_c have to be estimated for each haptic device such that the system
remains stable. The stability condition is derived from the transfer function of the
virtual coupling. Instead of finding the derivative, a finite difference estimates the
time discretised form of the spring-damper transfer function. One obtains

$$Z_C(z) = \left(b_c + \frac{k_c}{s} \right) \bigg|_{s \to \frac{1}{T}(1-z^{-1})} \tag{3.24}$$

which yields for the two-port of the virtual coupling

$$\begin{pmatrix} F_d \\ -v_e \end{pmatrix} = \begin{pmatrix} 0 & 1 \\ -1 & \frac{1}{Z_c(z)} \end{pmatrix} \begin{pmatrix} v_e \\ F_e \end{pmatrix} \tag{3.25}$$

Combining (3.22) and (3.25) the resulting matrix is

$$\begin{pmatrix} F_d \\ -v_e \end{pmatrix} = \begin{pmatrix} Z_d(z) & ZOH(z) \\ -1 & \frac{1}{Z_c(z)} \end{pmatrix} \begin{pmatrix} v_h \\ F_e \end{pmatrix} \tag{3.26}$$

The observation of the eigenvalues and poles lead to the following stability condition

$$Re\left(\frac{1}{Z_c(z)}\right) \geq \frac{1 - \cos(\varphi(ZOH(z)))}{2Re(Z_d(z))}|ZOH(z)|. \qquad (3.27)$$

The constants on the right hand side are either known or are determinable by device measurements. The constants on the left hand have to be chosen to satisfy the condition (3.27). Consequently the left hand term plotted in frequency domain should always be below the right hand term. For an evaluation of the virtual coupling with the *Omni* haptic device refer to [7]. Lee et al. [24] found out that the coupling produces small vibrations in the equilibrium, in their improvement they propose to replace the linear with a non-linear spring.

3.2 GRAB Device

In this chapter, a force-feedback device is presented which is used in the developed VR system. The haptic device named GRAB is a special construction conceived to provide force-feedback for index finger and thumb by two contact points. The VR system presented in this work is optimised for the use of this device as it provides the better interaction possibilities with the textiles compared to other devices. Since it has two end-effectors, it allows the grasping of objects and in the designated use-case of the VR system the assessment of textile properties. Although the device was designed allowing the aforementioned interactions with textiles (see Bergamasco et al. [5]), its initial development was made within an earlier EU-funded project with a different purpose, namely the access to graphics for blind people as shown in Iglesias et al. [20]. Figures 3.14 and 3.15 shows the complete setup of the device including the modifications for tactile feedback.

From the technical point of view the GRAB is a two finger (thumb and index finger) haptic device and consists of two mechanical arms. The arms with a telescopic mechanism are mounted to a grounded plate. These plates constitute the foundation for the force generators, the actuators, controlling the rotation, yaw and length of the arms. A separate amplifier box generates the needed power output for the motors transmitting the resulting torques by cables onto the joints of the arms. Optical encoders attached to the motors measure the resulting movement. This mechanism allows a force output in three degrees of freedom to the user. The positional force is perceived at the finger by thimbal end-effector at each arm. In our modified version, the end effector is replaced by a gimbal being a pivoted support in three axes with a small plate at the end. The gimbal is also at its joints sensorised to additionally measure the orientation of the finger. For stimulation of the tactile sense, the fingers are placed onto the plate with perpendicular vibrating pins coming out from a tactile array mounted on the backside. A force sensor connected to the plate measures the force applied on the finger by four strain gauges. The data coming from the rotation sensors of the gimbal and the strain gauges is sampled by a microcontroller. The values are sent via a conventional RS-232 serial interface to a control PC at a

Fig. 3.14 GRAB device, (below) amplifier (left) and control PC (right)

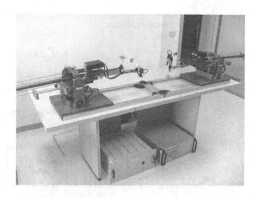

Fig. 3.15 Gimbal with tactile array and force sensor plate

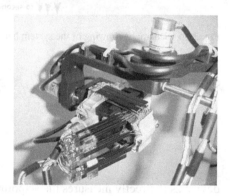

rate of 900 force and 100 rotation packets per second. This control PC processes measurements of the sensors and encoders to transform the signals into a position, orientation, and force information. The information is used to generate the analog output for the motors of the arms. The complete signal processing chain, the control loop, is running at approx. 5.2 kHz.

Communicating forces to and positional data from the device is managed by an Ethernet network connection. The computer on that the VR system is located continuously sends and receives UDP network packets containing the force respectively position and orientation vectors for both fingers. The transmission rate of these packets is variable but is usually at the rate of the control loop. A minimum rate of 240 Hz guards the haptic device against software crashes at high forces. The control PC shuts the motors down if packets are received in a lower rate.

The immediate adjustment of the motor current to get the correct force output is realised in the control loop in two stages as shown in Fig. 3.17. In the first, the open velocity loop, a current is estimated based on the motion of the end effector indirectly obtained from the optical encoders. As the mechanical parameters (e.g. inertia of end effector, friction) are known the output can be corrected to reduce these unwanted forces. In the second stage, the closed force control loop uses the data from the force sensor to correct the current with respect to the desired force on the end effector. The latter stage enhances the precision in the force output of the

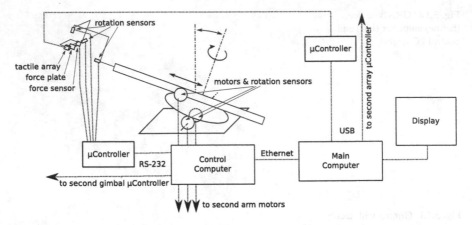

Fig. 3.16 Schematic drawing of the system hardware

Fig. 3.17 Control loop of the
GRAB (cf. [6])

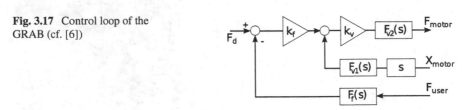

device as it directly measures the setpoint contrary to the velocity estimation of the
open loop obtained from a positional signal.

3.3 Haptic Rendering

The task in a VR system creating the touch perception is also often referred to as
haptic rendering. The term is related to the graphical rendering which provides visual information by rendering a scenery pixel by pixel. Despite the similarity of the
haptic rendering in providing a sensory *image*, it differs significantly in its computation.

The rendering of graphical images and thus the estimation of a pixel colours is
a local problem independent from pixel to pixel whereas the haptic rendering needs
to consider the complete geometry or even the physical state of the scenery to render a touch force. Therefore the demands to the computer system are by definition
much higher than to graphics. Considering further the stability aspect and the perception, the system should constantly deliver the results at an update rate of 1 kHz.
Techniques able to deliver force updates at these rates are called **direct rendering**
methods. In other cases where the rendering exhibits a high computational complexity, the haptic rendering makes use of the previously illustrated concept of virtual
coupling to stabilise the interaction. In the latter case, the actual model update rate
is some orders below 1 kHz. The ideal approach in rendering with virtual coupling

Fig. 3.18 Mass-spring system modelling finger contact

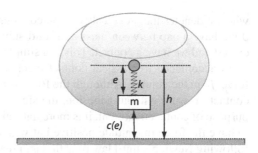

is to find the limits of the VE stiffness with respect to the local linear model and the haptic device characteristics such that the internal friction and other effects of the device dissipate the energy induced by sampling and conversion.

Independent of the rendering approach used, the rendering should resemble the touch feedback as far as possible. Mahvash et al. [25] coined the term **high-fidelity** regarding the haptic rendering. He specified properties which the system should meet the demands of high-fidelity haptic rendering. The properties are:

- the resemblance of virtual force response with actual responses
- force continuity under all interaction movements
- passivity of the virtual environment
- high force update to combat the adverse effects of discretisation

3.3.1 Contact Modeling

In both rendering methods, the requirement for rendering of contact forces is the detection of collisions between participating bodies within the physical simulation. Without it, simulations would not handle intersections and self-intersections properly. Since penetrations do not happen in reality a configuration with intersection is considered to be invalid and would obviously have a disturbing effect on the physical authenticity of the virtual reality.

The application of the contact forces in the physical simulation by considering the contact alone is quite difficult as the contact yields a highly non smooth displacement function and changes the system matrix. For finding the solution to the physical problem, it is therefore advantageous to consider the contact by constraints which provides a natural way to incorporate the contact conditions. Hence, mathematical tools to introduce constraints to the physical simulation will be provided.

As a simple example, a mass-spring system as shown in Fig. 3.18 is considered. The final state of the system has already been computed at no contact. Assuming now the spring is fixed above a rigid support at a distance of h apart from the attached mass.

The setup restricts the motion of the mass by

$$c(e) = h - e \geq 0 \tag{3.28}$$

where e denotes the previously computed elongation of the spring. In case of $c(e) > 0$ we have a gap between mass and rigid support, otherwise if $c(e) = 0$ the gap is closed and we have a contact. This is a simple example of a **gap function**.

In accordance to classical contact mechanics, the mass experiences a reaction force f_r in the contact. Although the force is assumed to be negative to account for contact pressure without adhesion, the sign is just a matter of convention. For our purpose of contact treatment, it is more natural in the context of Signorini's problem to have the force magnitude positive but with opposite direction. Consequently, by following Newton's third law of action and reaction, the constraint is defined by

$$f_r \geq 0 \tag{3.29}$$

It should be noted that the contact problem can enter two states. The first is reached when the spring stiffness is large enough to avoid the contact with the rigid support. In this case, the following conditions hold true

$$c(e) \geq 0 \quad \wedge \quad f_r = 0 \tag{3.30}$$

And one arrives at the second state when the mass touches the support. Then, the conditions

$$c(e) = 0 \quad \wedge \quad f_r \geq 0 \tag{3.31}$$

hold. It can therefore be concluded

$$c(e) \geq 0 \quad \wedge \quad f_R \geq 0 \quad \wedge \quad c(e) \cdot f_r = 0 \tag{3.32}$$

These are known as the contact conditions for Signorini's problem and will be used to solve the problem under the terms of contact. Unfortunately, application of the conditions would introduce variational inequalities in our variational statement of the system's energy which can't be regarded in the approach. Methods have been proposed to either resolve the inequalities or reformulate the problem suitable for the variational approach.

One well-known method is the **Lagrange multiplier method**. The method adds to a constraint term to the energy of the system to consider the conditions of the contact problem. For each condition a term is added, allowing to account for various effects which can be expressed in terms of energy. In our special case, it yields a single term $\lambda c(e)$

$$\Pi(e, \lambda) = \Pi(e) + \lambda c(e) \tag{3.33}$$

where λ is called **Lagrange multiplier** and is real valued. Substituting the general energy function with the strain energy of the spring, one obtains

$$\Pi(e, \lambda) = \frac{1}{2}ke^2 - mge + \lambda c(e) \tag{3.34}$$

as energy of the system. In our particular case, the Lagrange multiplier equals the reaction force f_r. By varying δe and $\delta \lambda$ independently, the variation of the energy leads to the two equations

$$ke\delta e - mg\delta e - \lambda e = 0 \tag{3.35}$$

$$c(e)\delta\lambda = 0 \tag{3.36}$$

The first equation accounts for the equilibrium of the system including the contact force at the touch of the rigid support, and the second ensures the fulfilment of the kinematic constraint for the contact. Solving for the multiplier yields

$$\lambda = kh - mg = f_r \tag{3.37}$$

Despite the expression of the contact with Lagrange multipliers, it is still necessary to check for contact by $f_r > 0$. When no contact is detected, i.e. f_r equals zero, then the inequality constraint is disregarded as the multiplier vanishes.

Another common method to consider contact problems is the **penalty method** which also adds a term to the energy. But instead of strictly enforcing the geometric condition $c(e) = 0$ at contact, it relates a penalty term to the penetration of the contacted surface. In the mass-spring system it leads to

$$\Pi_p(e) = \Pi(e) + \frac{1}{2}\varepsilon[c(e)]^2 \quad \text{with } \varepsilon \geq 0 \tag{3.38}$$

with ε as penalty factor. Inserting our problem gives

$$\Pi(e\lambda) = \frac{1}{2}ke^2 - mge + \frac{1}{2}\varepsilon[c(e)]^2 \tag{3.39}$$

The term can be seen as the energy of a spring that is compressed due to interpenetration of the bodies. The variational approach leads to

$$ke\delta e - mg\delta e - \varepsilon c(e)\delta\varepsilon = 0 \tag{3.40}$$

with its solution

$$e = \frac{mg + \varepsilon h}{k + \varepsilon} \tag{3.41}$$

In the contact, for the constrained problem

$$c(e) = h - e = \frac{kh - mg}{k + \varepsilon} \tag{3.42}$$

a penetration of the rigid support by the mass occurs since $mg \geq kh$. It can be observed further that the amount of penetration depends on the choice of ε. For the limit $\varepsilon \to \infty$, the geometric constraint is ensured but large stiffness can lead to numerical instabilities. Thus, the value should be set appropriately as a compromise between physical correctness and stability.

The most versatile method is given by the **Signorini's problem** as follows. To elaborate the idea of the method, a deformable body given by the closure of a set

Fig. 3.19 Signorini's contact problem

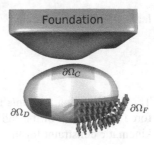

Ω of particles in $\mathbb{R}^3$ is considered. It is defined as reference configuration of the body at time $t = 0$ in Cartesian coordinates. Figure 3.19 shows an example of a configuration.

The body comes into contact with a rigid support or foundation which constrains the motion of the body in one direction. Let $\partial\Omega_C$ denote the contacting part of the body's surface. A gap function defined by g giving the distance between the body and the foundation in movement direction for all points in $\partial\Omega_C$. As before, it follows from the kinematics that

$$g \geq 0 \tag{3.43}$$

Moreover, for the physical problem, the condition must be in accordance with the state of stresses on the contact surface. Let denote σ Cauchy's stress tensor at material coordinate $\mathbf{X}$. Further let $\mathbf{x}$ be the position of $\mathbf{X}$ in the deformed configuration along with the surface normal $\mathbf{n}$ at that point. By making use of Cauchy's formula (2.46), the surface stress can be decomposed into tangential and normal components with

$$\sigma^N(\mathbf{x}) = \sigma_{ij}\mathbf{n}_i(x)\mathbf{n}_j(x) \tag{3.44}$$

$$\sigma_i^T(\mathbf{x}) = \sigma_{ij}\mathbf{n}_j(x) - \sigma^N\mathbf{n}_i(x) \tag{3.45}$$

Based on the decomposition, it can be stated that the normal stress equals zero at no contact and yield the second contact condition

$$\sigma^N(\mathbf{x}) = 0 \quad \Longleftrightarrow \quad g(\mathbf{x}) \geq 0 \tag{3.46}$$

These conditions are defining the three constraints as of (3.32) which are in general form

$$g(\mathbf{x}) \geq 0 \perp \sigma^N(\mathbf{x}) \geq 0 \tag{3.47}$$

For further consideration, the body will be subjected to body forces $\mathbf{f}$ and surface traction $\mathbf{t}$ on a part $\partial\Omega_F$. The fixation of a portion $\partial\Omega_D$ of $\partial\Omega$ is also allowed. Additionally, for simplicity, the effects of friction are neglected by letting tangential stresses vanish in the contact area. The investigations are limited to infinitely small deformations $\mathbf{u} = x - \mathbf{X}$ of the body. Regarding these conditions, one comes to the set of equations

$$\nabla \sigma = \mathbf{f} \quad \text{for all } \mathbf{x} \in \Omega \qquad \text{(equilibrium condition)}$$

$$\mathbf{u} = 0 \quad \text{for all } \mathbf{x} \in \partial \Omega_D \qquad \text{(geometric constraint)}$$

$$\sigma_{ij} \mathbf{n}_j = \mathbf{t}_i \quad \text{for all } \mathbf{x} \in \partial \Omega_F$$

$$\sigma_i^T = 0 \qquad \text{(frictionless contact)}$$

$$g(\mathbf{x}) \geq 0 \quad \perp \quad \sigma^N \geq 0 \qquad \text{(contact constraint)}$$

$$(3.48)$$

This set of equation defines Signorini's problem in its traditional form.

The calculations needed to solve the Signorini's problem numerically for haptic interaction are elaborated in the following. The approach shown here was introduced by Duriez et al. in [12, 13]. As the contact problem given by the conditions is related to quadratic programming, a branch of optimisation theory, it can be formulated as a linear complementary problem (LCP). The LCP is a special case of quadratic programming which seeks to find two vectors $\mathbf{z}$ and $\mathbf{w}$ solving a matrix equation of the form

$$\mathbf{w} - M\mathbf{z} = \mathbf{q} \qquad (3.49)$$

In the LCP, the matrix M and $\mathbf{q}$ are known. The vectors $\mathbf{z}$ and $\mathbf{w}$ are governed by inequality and complementary constraints as follows

$$\mathbf{w} \geq 0 \qquad (3.50)$$

$$\mathbf{z} \geq 0 \qquad (3.51)$$

$$w_j z_j = 0 \quad \text{for all } j \qquad (3.52)$$

where the inequality is treated component wise (cf. [27]).

In the contact problem, contact forces and resulting displacements are the unknowns of the problem and fulfil the conditions as stated above. For the linearised search, the problem is fixed in contact space to constant gap function and contact normals at each time step. Thus, the displacements are constraint to normal direction.

To find the contact points, the approach lets the involved bodies initially move without the contact constraints. At an interpenetration, the contact is expressed in the gap g along with the contact point and its barycentric coordinates in the element to linearly relate the gap to the displacement of the element nodes (see Fig. 3.20). In an edge to edge contact one has for example the mapping of the displacements in contact space given by

$$g = \mathbf{n}^T \left([\alpha_1 \ \beta_1] \begin{bmatrix} \mathbf{U}_1 \\ \mathbf{V}_1 \end{bmatrix} - [\alpha_2 \ \beta_2] \begin{bmatrix} \mathbf{U}_2 \\ \mathbf{V}_2 \end{bmatrix} \right) + g^{free} \qquad (3.53)$$

where α_i and β_i are the barycentric coordinates of the contact point in body i and $\mathbf{U}_i, \mathbf{V}_i$ the displacements of the nodes to which the edge belongs. n^T is the trans-

Fig. 3.20 Contact space after
collision of two elements

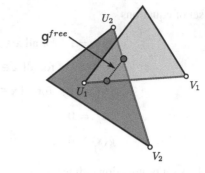

Fig. 3.21 Possible contact
cases between two triangles

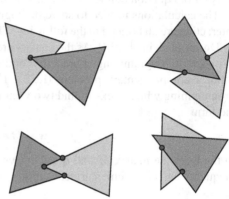

posed unit normal of the contact, and g^{free} the gap between the bodies in uncon-
strained motion. The displacements occurring at point to face contact is given accordingly by

$$g = \mathbf{n}^T \left([\alpha_1 \ \beta_1 \ \gamma_1] \begin{bmatrix} \mathbf{U}_1 \\ \mathbf{V}_1 \\ \mathbf{W}_1 \end{bmatrix} - [\alpha_2][\mathbf{U}_2] \right) + g^{free} \tag{3.54}$$

In general the possible contact configurations are as illustrated in Fig. 3.21. One can
combine all contact mappings into a transfer matrix such that the gap is given by

$$\mathbf{g} = [H_1] \begin{bmatrix} \mathbf{U}_1 \\ \mathbf{V}_1 \\ \vdots \end{bmatrix} - [H_2] \begin{bmatrix} \mathbf{U}_2 \\ \mathbf{V}2 \\ \vdots \end{bmatrix} + \mathbf{g}^{free} \tag{3.55}$$

The resulting contact forces $\mathbf{F}_i^C$ are related to the forces $\mathbf{F}_i$ on the nodes with the
same transfer matrix H such that

$$\mathbf{F}_i = H_i^T \mathbf{F}_i^C \tag{3.56}$$

yields the forces of each object.

The Signorini's problem is described as a linear complementary problem. It is already known that the inequality conditions of the problem are identical to those of the LCP. And with the formulation of the contact space one has partly succeeded in linearising our initial problem which is needed for the application of LCP solution methods. For the complete linearisation, it is needed to express the physical problem in a stiffness term by

$$K\mathbf{u} = \mathbf{f} \tag{3.57}$$

to obtain a linear relationship to the displacements. The linearised problem is finally solvable by several computational methods as given in [27].

With the compliance matrix being $C = K^{-1}$, one can relate the position of free motion $\mathbf{P}^{free}$ to the contact forces $\mathbf{F}^C$ and the constrained position $\mathbf{P}^C$:

$$\mathbf{P}^C = C\mathbf{F}^C + \mathbf{P}^{free} \tag{3.58}$$

Inserting the relationship into (3.55), one gets for the gap

$$\mathbf{g} = (H_1 C_1 H_1^T + H_2 C_2 H_2^T)\mathbf{f} + \mathbf{g}^{free} \leq 0 \tag{3.59}$$

and for the associated LCP

$$\mathbf{f} \geq 0 \tag{3.60}$$

$$\mathbf{g} = (H_1 C_1 H_1^T + H_2 C_2 H_2^T)\mathbf{f} + \mathbf{g}^{free} \leq 0 \tag{3.61}$$

$$\mathbf{f} \perp \mathbf{g} \tag{3.62}$$

If the resulting matrix of the LCP is symmetric positive definite, the LCP can be solved by a principal pivoting method and Lemke's algorithm for co-positive matrix. For details about this method refer to [27].

With the established relation of the Signorini's problem to the LCP, the solution to the contact constraint given by (3.32) can be found. However, in the formulation, the frictional effects are disregarded by setting tangential stresses to zero. But for grasping and pulling textiles, it is crucial to employ such effects. To consider the frictional effects of the interaction, the laws of Amonton and Coulomb are appropriate for the simulation. They codified their findings in three laws being the proportionality of friction force to the applied load, the independency of contact surface wrt. friction force, and the independency of friction force wrt. to sliding velocity. The Coulomb friction model regards the laws with an additional differentiation between stick and slip contact. The stick condition determines the state of contact which is given by the relation

$$F^T \leq \mu_s F^N \tag{3.63}$$

where F^T is the applied tangential and F^N respectively the normal force and μ_s a dimensionless quantity, the static friction coefficient. When the tangential force exceeds the limit defined by the condition, the state changes from initially stick to sliding contact and results in a drop of the opposing friction force to the dynamic or kinetic friction given by

$$F^f = \mu_d F^N \tag{3.64}$$

where F^f is the resulting friction force and μ_d the dynamic friction coefficient which is usually lower than the coefficient for static friction. Due to the simplicity of the presented model, these laws will be introduced in the formalism. For preserving the compatibility to the contact space, an expression of the friction in terms of tangential displacement is needed:

$$\mathbf{g}_t = 0 \quad \Longrightarrow \quad \|\mathbf{f}_t\| < \mu_s \|\mathbf{f}_n\| \qquad \text{(stick condition)}$$

$$\mathbf{g}_t \neq 0 \quad \Longrightarrow \quad \mathbf{f}_t = -\mu_d \|\mathbf{f}_n\| \frac{\mathbf{g}_t}{\|\mathbf{g}_t\|} \qquad \text{(slip condition)}$$

Unfortunately, the non-linearities introduced by the condition and state change complicates the problem. This makes the direct evaluation by means of an LCP impossible. The computation of friction can also not be separated from the contact resolution as the tangential forces at every node might have influence on the friction state of the others. For one contact point, a fast and precise formulation has been proposed as for example in [25]. But as the textile is highly deformable, during interaction it is most-likely that multiple contact points are apparent. Hence, the latter method is not applicable.

For simplification of the problem, the common approach is to evaluate both conditions by an approximation scheme which is called friction cone. Instead of taking the vector norms, the stick condition is met when the vector sum of normal and tangential force lies within this cone. To integrate the friction cone technique into the LCP, an k-sided pyramid is commonly used to get a linear approximation of the cone. Hence, the new LCP is given by

$$0 \quad \leq g_n \quad \perp \quad f_n \geq 0 \tag{3.65}$$

$$0 \quad \leq \mu + g_{t_1} \quad \perp \quad f_{t_1} \geq 0 \tag{3.66}$$

$$\cdots$$

$$0 \quad \leq \mu + g_{t_k} \quad \perp \quad f_{t_k} \geq 0 \tag{3.67}$$

$$0 \quad \leq \mu f_n - \sum_{r=1}^{k} f_{t_r} \quad \perp \quad \mu \geq 0 \tag{3.68}$$

where μ is not a physical quantity but represents the sliding displacement at the contact (cf. [3]). The extension made to the LCP however creates a considerable bigger system matrix, e.g. an appropriate approximation of the friction cone by eight sides yields a ten times larger matrix. As an alternative approach, Duriez et al. [13] proposed a method using a linearised system behaviour and solve for each contact point while leaving the contributions of the other points constant. For an analysis of the existence and uniqueness of the solution to Signorini's problem refer to [8].

Irrespective of having friction included or not, for a proper contact handling it is necessary to detect the contact points originated from the intersections of two bodies in free motion. These points as shown before impose physical constraints to the simulation of the bodies such that the collision is resolved before coming visible to the user. For physical plausibility and interaction, collision handling is an important task for virtual reality systems. Typically, it is separated into two categories, collision detection and resolution. While the resolution was partly considered in the previous Section, the methods of collision detection and its application in the context of the VR system are now discussed.

3.4 Collision Detection

Since the detection of collisions is required in many applications dealing with object interaction especially with a user controlled object or physical bodies, the topic is a extensively researched topic. The underlying algorithms are further differentiated into continuous and discrete collision detection.

In the discrete case the detection algorithm checks if collisions are present at a time step respectively geometrical fixed configuration. Hereby, imminent collisions are ignored since motions or physically motivated changes in geometry are not considered. In situations of thin or fast objects in physical simulations, the detection is likely to oversee collisions taking place between two time steps. This is not caused by a wrong or inaccurate analysis of the current configuration but of the skipped time step in between where the intersection has occurred.

The other category being the continuous detection algorithms are reliable in such situations having a physically motivated change of the geometry. These algorithms not only detect a collision but also know the time step at which the collision has occurred. The reliability of this kind of detection algorithms is bought with increased complexity in the search of intersection. But the drawback is marginal in comparison to the reliability of the discrete collision detection.

But for all detection algorithms of any category, it is needed to test every object against the other, which is already of $\mathcal{O}(N^2)$ complexity in run-time. Depending on the primitives of which the object is made up of, e.g. triangle, edges, vertices, the aforementioned number increases even more as detection against each primitive type has to be tested. Therefore, in real time applications the global test is unsuitable due its high demands. Typically, detection algorithms make use of topology and locality of the object's structure. By splitting the search in a broad phase considering only rough approximation of the objects boundary and narrow phase on the primitive level, algorithms gain in performance.

3.4.1 Bounding Volumes

The convex approximation of the objects shape in detection algorithms are called bounding volumes (BV). These volumes are used to describe the spatial extension

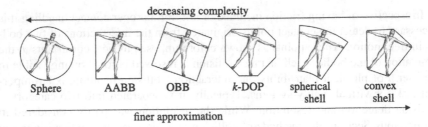

Fig. 3.22 Correlation between approximation and efficiency of BV

of the object in such manner that the interaction query is reliable and efficient. The reliability in this context means that an intersection of two objects is present if and only if the BVs have intersected.

Although the properties of the BV are defined, its shape can still be arbitrary. Depending on the algorithm and its application, the shape can vary from coarse to fine approximation, but the limiting factor remains the efficiency in the query. Figure 3.22 illustrates the trade-off between accuracy of the BV and complexity in the query.

In the following, the estimation of the BV representations will be outlined with respect to the geometry of the object. Without loss of generality, it is assumed that an object O consists of a set of points $\mathbf{p}_i \in \mathbb{R}^3$. The order on points respectively vectors is defined as component wise order of the real numbers.

3.4.1.1 Bounding Sphere and Bounding Box

The bounding sphere (BS) and the (axis-aligned) bounding box (AABB) are certainly the simplest BV. The first BV is solely defined as the circumsphere of the point set, whereas the bounding box is given by the maxima and minima within the point set. Thus for the underlying point set, the volumes are determined by the conditions

$$\forall \mathbf{p}_i \in O : (\mathbf{p}_i - \mathbf{c})^2 \leq r^2, \qquad\qquad BS\{\mathbf{c}, r\} : \text{center and radius}$$
$$\forall \mathbf{p}_i \in O : \mathbf{p}_i \geq \mathbf{p}_{min} \wedge \mathbf{p}_i \leq \mathbf{p}_{max}, \qquad AABB\{\mathbf{p}_{min}, \mathbf{p}_{max}\} : \text{lower and upper edge}$$

whereas the bounding sphere is invariant with respect to object rotation.

3.4.1.2 Oriented Bounding Box and Discrete Oriented Polytope

The so-called oriented bounding box (OBB) is an extended definition of the aforementioned bounding box. Here, the alignment is not restricted to Cartesian coordinates. The additional storing of an orthonormal basis along with its origin to the OBB allows in principle the definition of arbitrary rotated AABB. Although the

OBB shares the common idea of the bounding box, its computation is more complicated if a reasonable orientation should be chosen. The computation of the orientation is similar to the discrete oriented polytope and its concept will be illustrated in the next section. Nonetheless the main advantage of the OBB is the invariance to rotations.

3.4.1.3 Discrete Oriented Polytope

A discrete oriented polytope or in short, k-DOP, adds more flexibility in the definition of the bounding volume. The bounding of geometry is given by arbitrary half-spaces defined by normals $\mathbf{n}_i$, with different orientations $i = 1, \ldots, k/2$. To each normal $\mathbf{n}_i$ are two scalars associated describing the minimal d_i^{min} and maximal d_i^{max} distance of the object from the halfspace. In total, a triple $(\mathbf{n}_i, d_i^{min}, d_i^{max})$ describes the bounding planes BP

$$BP_i^{min}: \quad \mathbf{n}_i \mathbf{x} + d_i^{min} = 0$$

$$BP_i^{max}: \quad \mathbf{n}_i \mathbf{x} + d_i^{max} = 0$$

which create a so-called Slab S_i. The final convex polytope consist of the intersection of all slabs S_i.

$$k\text{-}DOP: \quad \bigcap_{1 \leq i \leq k/2} S_i$$

Figure 3.24(a) gives an example of an 8-DOP. The k-DOP BV was firstly used by ray-tracing algorithms finding efficiently ray-triangle intersections and was introduced by [21]. Commonly used values for k-DOPs are $6, 10, 14$ an 26, whereas a 6-DOP is equal to AABB. Thus, the k-DOP shares the property of being variant to rotations as Fig. 3.24(b) illustrates. Another property of a DOP is that it converges for $k \rightarrow \infty$ to the convex hull of the object.

An important factor for the quality of the object fitting using k-DOP is the choice of the normals. The simplest and easiest choice is to have the normals fixed for any k-DOP. By having only one set of normals for all objects the intersection test is simplified to an interval overlap check. In this aligned case, an intersection of the BVs is present if only one of these intervals overlaps.

The disadvantage of a common set of normals is the fitting behaviour of the k-DOP. Since it does not consider the shape of the object, the fitting of BV might not be optimal. As Fig. 3.24(b) shows, the bounding volume created by the initial set of normals (in grey) is bigger than the new one (in black). But allowing different sets in the DOPs makes it necessary to transform the minimum and maximum values for a comparison between two objects. In [34] a scheme for transforming one DOP to the basis of the other is proposed in order to compare the distance values. With the knowledge of the affine transformation M mapping the DOPs on each other and the normals defining the half-spaces, a inverse mapping can be found to transform the distances d_i^{min}, d_i^{max} in the other objects space.

Fig. 3.23 Schematic drawing
of an 8-DOP

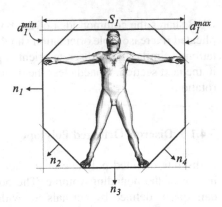

Fig. 3.24 Rotated object
including the DOP (*grey*) and
the recomputed DOP (*black*)

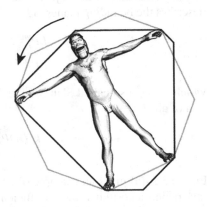

3.4.1.4 Spherical Shell

The bounding volume represented by a spherical shell was first introduced by [22]
and is defined as a volumetric portion between two concentric spheres. Its basic idea
is to achieve a better approximation of the BV by exploiting the objects curvature.
The complete description based on the aforementioned BV satisfies the following
conditions for each point

$$r_{inner}^2 \leq \mathbf{p}_i^2 \leq r_{outer}^2$$

$$\frac{\mathbf{p}_i \, \mathbf{a}}{\|\mathbf{p}_i\|} \geq \cos(\alpha)$$

Whereas the latter condition ensures the points to lie within a cone given vector **a**
and a angle α. The volumetric portion created by fulfilling both condition is illustrated in Fig. 3.25.

Fig. 3.25 Definition of a spherical shell in $\mathbb{R}^2$

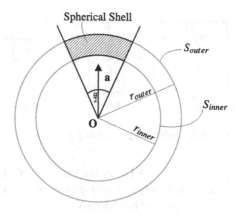

3.4.1.5 Convex Hull

The aforesaid BV give only an approximate convex hull of the object, but it is also possible to use directly the convex hull as a BV. With the criterion of the convex set containing a set of points and their connecting lines, the convex hull is defined as the minimal convex set with respect to a set of points. Although the convex hull is the best convex BV of the object, the overlap test and the update is too costly. Hence, its use for collision detection is very limited.

3.4.2 Bounding Volume Hierarchy

The introduced bounding volumes allow a more efficiently testing of object overlaps prior the costly intersection tests. But if the BVs of two objects do overlap it is not necessarily the case for the objects. For the clear determination and proper collision handling it is still needed to know if and which primitives do overlap and other properties like penetration depth. These test still require a lot of computation time by testing against all primitives.

To reduce the complexity in the check by removing primitives not involved in the overlap, a tree hierarchy is added on top of the object called bounding volume hierarchy (BVH). With the BV being the root node of a tree, the underlying object is split into two (or more) smaller parts with a BV at each child node. The splitting takes place until a stop criterion has been reached, e.g. minimum number of primitives (see [23] for more). In best case the recursive splitting will consider the geometric extensions of the object.

The nodes of a BVH are as common for tree structures distinguished between internal and external (*leaf*) nodes. The inner nodes contain only the references to their children and the BV for the corresponding split part of the object. Leaves on the contrary contain concrete information to the object as they store sets of primitives and the associated BV, but no references to children. Figure 3.26 illustrates the

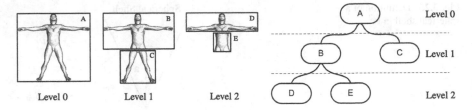

Fig. 3.26 Split of object creating a BVH

structure of a BVH. By the hierarchical structure of the object, the detection algorithm needs only to traverse the tree down to the leaves reducing the complexity in the average case to $\mathcal{O}(\log(N))$.

3.4.2.1 Construction

Since the hierarchy is constructed in a tree-like manner, the BVH inherently possesses the properties of a tree. These properties namely the depth of the BVH, branching factor and the leaves size have great impact on the performance of the collision query. For instance, an inappropriate split of the object would lead to an unbalanced tree having a high depth and thus long traversals to the leaves. Contrary to high depths a relatively flat tree would lead to leaves with a high amount of primitives yielding the benefit of BVH ineffective for collision test. Thus, the construction of the BVH has a significant influence on the final performance of the collision detection.

There are two approaches in constructing the BVH, the bottom-up and the top-down approach. Although both approaches following the conflicting goals of having a flat tree and small leaf sizes, they differ in the starting point. The bottom-up starts at the primitive level of an object and merges these primitives according to a shape criterion, e.g. curvature, to create leaves of the BVH. With the use of such criterion, regions of similar shapes are aggregated. Hence, the bottom-up construction is often used for self-collision detection separating flat areas from curved areas.

The second, the top-down approach, is more general and suits better for inter object collision detection being the main application in haptic rendering. The starting point here is the whole object. As already indicated before, in an recursive process the object is separated into sub-objects. Depending on the branching factor, one or more split planes define to which sub-objects primitives belong to. Since the split plane can be place arbitrarily in space, it has influence on the size of the sub-objects. For a good placement of the plane, two methods exists, the **longest-side-split** and the **splatter-split** as illustrated below in Fig. 3.27. The first and more simple split aligns the normal of the split plane along one coordinate axis in which the object has the widest extension S_{max}. The center of gravity with respect the primitive type defines then the placement of the plane in the chosen coordinate direction. Depending on the branching factor, more iterations of the previous steps on the other axes are performed. The splatter-split tries to separate on the axis at which object exhibits

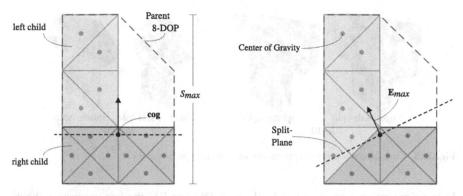

Fig. 3.27 Comparison of *longest-side* and *splatter*-split

the highest dispersion. For that purpose the method computes from all centers of gravity of the primitives a covariance matrix with respect to their coordinates. By estimating the eigenvector $\mathbf{E}_{max}$ with biggest eigenvalue of the matrix the direction of the split normal is found. The placement of the split plane is then also defined by the center of gravity of all primitives denoted by $\mathbf{cog}_i$.

Thus, the covariance of one random variable is calculated with

$$Cov(X, X) = Var(X) = \frac{\sum_{i=1}^{n} cog_i^x - \overline{cog}^x}{n}$$

and for different variables

$$Cov(X, Y) = Cov(Y, X) = \frac{\sum_{i=1}^{n}(cog_i^x - \overline{cog}^x)(cog_i^y - \overline{cog}^y)}{n}$$

Yielding the covariance matrix K as follows

$$K = \begin{pmatrix} Var(X) & Cov(X, Y) & Cov(X, Z) \\ Cov(X, Y) & Var(Y) & Cov(Y, Z) \\ Cov(Y, Z) & Cov(Y, Z) & Var(Z) \end{pmatrix}$$

As the covariance matrix is symmetric, the eigenvalues are only real valued. The eigenvalue and its corresponding eigenvector indicate in the split case, the direction of highest dispersion which leads to a sub-objects of nearly equal sizes.

3.4.3 Complete Collision Query

The initial simple test of intersecting primitives of two objects O_1 and O_2 has been replaced by an improved algorithm working as follows. Assuming the associated BVH for O_1 and O_2 have already been constructed, the test begins to compare the root nodes of both BVH. Hereby, the corresponding BVs are tested for overlapping.

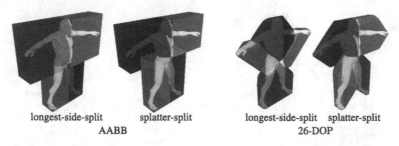

longest-side-split   splatter-split   longest-side-split   splatter-split
                AABB   26-DOP

Fig. 3.28 Comparison top-down approaches for construction of the BVH

In the simplest case of having AABB and k-DOP as BV, the test requires a check of the intervals created by the min-/max-values on the axes. For other or mixed BV types, a simple overlap test has to compare edges of O_1 against face O_2 and vice versa leading to 144 test for a OBB structure. A reduction can achieved as proposed in [15], by the use of the separation axis theorem. It states that two convex polytopes are disjoint iff there exists a separating axis being orthogonal to a face either polytopes or to an edge from each polytope. In their case of an OBB structure, the compare reduces to 15 checks of possible separating axes in the worst case.

On positive overlapping, the algorithm continues to decent in the tree to the children of BVH(O_1) and BVH(O_2) and the previous step is repeated. These steps are iterated as long as there is an overlap detected and both tree nodes are inner nodes. In the latter case, the test has reached the leaves which contain the reduced set of primitives of the objects. In the other case, the objects are not intersecting.

As the final step, an intersection test has to ensure if and where the object primitives intersect. Depending on the collision response different intersection tests are required, which will be explained in the next section. Since most applications favour triangles for the objects primitives, several algorithms have been developed to detect intersection between this type of primitives. The algorithms differ in reliability and computational speed. The fast and most popular algorithms because of their geometric view are [19] and [28]. The idea behind the latter approach is the treatment of the triangles within their local frame of reference. To illustrate the working principle of the algorithm see Fig. 3.29 as an example. At start, the algorithms tries to reject non-intersecting triangles. For doing this, it first uses the plane equation of triangle B to compute the relative distances d_0, d_1, d_2 of the points of triangle A. These distances determine the location of triangle A with respect to the plane of B. If the d_i's have the same sign and $d_i \neq 0$ then the triangle has to be rejected as it lies on one side of the plane. Otherwise, the test is repeated by using plane equation of triangle A. If the sign change occurs in both triangles then the intersection must lie on the line generated by $\mathbf{N}_0 \times \mathbf{N}_1$ as seen in Fig. 3.30. With the previously computed distances d_i and the projection $\bar{p}_i$ onto the intersection line of both triangles, an interval $[t_0, t_1]$ for each triangle can be estimated. By comparing the interval borders the intersection is determined. Furthermore, in order to detect co-planar configurations of the triangles or edges a check against an epsilon ($|d_i| < \varepsilon \Rightarrow d_i = 0$) is needed. In this case the intersection test is observed in 2D given by the shared plane.

Fig. 3.29 Spatial configuration

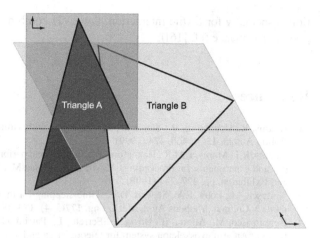

Fig. 3.30 Interval estimation

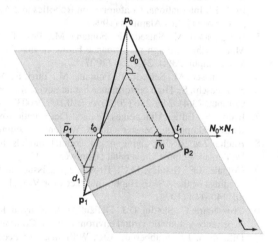

After showing up the different BV and construction methods, which are only an excerpt of the possible collision detection schemes, it raises the question which approach is the best. But this question cannot be answered without considering the specific application. In [15] a cost function (see (3.69) is given considering the number of potential operations (bounding volume N_{BV} and primitive check N_P) and its associated costs C_{BV}, C_P required to find collisions. The total cost T and thus the time for the detection is governed by the conflicting constraints of having minimal number of overlap checks by tight fitting of the BVH and a fast test for BV overlaps.

$$T = N_{BV}C_{BV} + N_PC_P \qquad (3.69)$$

The more the BV corresponds to the convex hull or the deeper the hierarchy is constructed, the higher are the costs and amount of the overlap test. Since these parameters depend on the geometry and their placement inside the scene, each case must be experimentally tested to find the optimal configuration. For most applications especially for textile interaction, the BVH combined with a k-DOP achieves good performance (cf. [16]).

References

1. Adams, R., Hannaford, B.: Stable haptic interaction with virtual environments. IEEE Trans. Robot. Autom. **15**(3), 465–474 (1999)
2. Adams, R.J., Moreyra, M.R., Hannaford, B.: Stability and performance of haptic displays: theory and experiments. In: Proceedings ASME International Mechanical Engineering Congress and Exhibition, pp. 227–234 (1998)
3. Anitescu, M., Potra, F.A., Stewart, D.E.: Time-stepping for three-dimensional rigid body dynamics. Comput. Methods Appl. Mech. Eng. **177**(3–4), 183–197 (1999)
4. Bergamasco, M., Allotta, B., Bosio, L., Ferretti, L., Parrini, G., Prisco, G., Salsedo, F., Sartini, G.: An arm exoskeleton system for teleoperation and virtual environments applications. In: IEEE International Conference on Robotics and Automation, pp. 1449–1449. IEEE Computer Society, Los Alamitos (1994)
5. Bergamasco, M., Salsedo, F., Fontana, M., Tarri, F., Avizzano, C., Frisoli, A., Ruffaldi, E., Marcheschi, S.: High performance haptic device for force rendering in textile exploration. Vis. Comput. **23**(4), 247–256 (2007)
6. Bergamasco, M., Salsedo, F., Fontana, M., Tarri, F., Avizzano, C.A., Frisoli, A., Ruffaldi, E., Marcheschi, S.: High performance haptic device for force rendering in textile exploration. Vis. Comput. **23**(4), 247–256 (2007). doi:10.1007/s00371-007-0103-1
7. Blume, S.: Entwicklung eines Realzeit-Deformationsmodells der Fingerkuppe zur haptischen Kraftrückkopplung. Master's thesis, Welfenlab, Leibniz Universität Hannover (2007)
8. Buchmann, R.: Improving finger contact models for haptic interaction. Master's thesis, Welfenlab, Leibniz Universität Hannover (2009)
9. Colgate, J.E., Stanley, M.C., Brown, J.M.: Issues in the haptic display of tool use. In: Proceedings of the ASME Haptic Interfaces for Virtual Environment and Teleoperator Systems, pp. 140–144 (1994)
10. Cruz-Neira, C., Sandin, D.J., DeFanti, T.A., Kenyon, R.V., Hart, J.C.: The CAVE: audio visual experience automatic virtual environment. In: Communications of the ACM (1992)
11. DiStefano, J.J., Stubberud, A.R., Williams, I.J.: Feedback and Control Systems. Schaum's Outline Series (1995)
12. Duriez, C., Andriot, C., Kheddar, A.: Signorini's contact model for deformable objects in haptic simulations. In: IEEE/RSJ International Conference on Intelligent Robots and Systems, 2004. (IROS 2004). Proceedings, vol. 4, pp. 3232–3237 (2004). doi:10.1109/IROS.2004.1389915
13. Duriez, C., Dubois, F., Kheddar, A., Andriot, C.: Realistic haptic rendering of interacting deformable objects in virtual environments. IEEE Trans. Vis. Comput. Graph. **12**(1), 36–47 (2006)
14. Gillespie, R.B., Cutkosky, M.R.: Stable user-specific haptic rendering of the virtual wall. In: Proceedings of the ASME International Mechanical Engineering Congress and Exhibition, vol. 58, pp. 397–406 (1996)
15. Gottschalk, S., Lin, M., Manocha, D.: OBBTree: a hierarchical structure for rapid interference detection. In: Proceedings of the 23rd Annual Conference on Computer Graphics and Interactive Techniques, pp. 171–180. ACM, New York (1996)
16. Hanel, M.R.: Untersuchung und implementation verschiedener kollisionserkennungsalgorithmen für deformierbare körper. Master's thesis, Welfenlab, Leibniz Universität Hannover (2007)
17. Hayward, V.: Toward a seven axis haptic device. In: Proc. Int. Conf. Intelligent Robots & Systems (1995)

18. Hayward, V., MacLean, K.E.: Do it yourself haptics. Part I. REd **3**(2), 12 (2007)
19. Held, M.: ERIT: a collection of efficient and reliable intersection tests. J. Graph. Tools **2**(4), 25–44 (1998)
20. Iglesias, R., Casado, S., Gutierrez, T., Barbero, J., Avizzano, C., Marcheschi, S., Bergamasco, M., Labein, F., Derio, S.: Computer graphics access for blind people through a haptic and audio virtual environment. In: The 3rd IEEE International Workshop on Haptic, Audio and Visual Environments and Their Applications, 2004. HAVE 2004. Proceedings, pp. 13–18 (2004)
21. Kay, T.L., Kajiya, J.T.: Ray tracing complex scenes. ACM SIGGRAPH Comput. Graph. **20**(4), 269–278 (1986)
22. Krishnan, S., Pattekar, A., Lin, M., Manocha, D.: Spherical shell: a higher order bounding volume for fast proximity queries. In: Robotics: The Algorithmic Perspective: 1998 Workshop on the Algorithmic Foundations of Robotics, pp. 177–190 (1998)
23. Langetepe, E., Zachmann, G.: Geometric Data Structures for Computer Graphics. AK Peters, Wellesley (2006)
24. Lee, M.H., Lee, D.Y.: Stability of haptic interface using nonlinear virtual coupling. In: IEEE International Conference on Systems, Man and Cybernetics, vol. 4 (2003)
25. Mahvash, M., Hayward, V.: High-fidelity haptic synthesis of contact with deformable bodies. IEEE Comput. Graph. Appl. **24**(2), 48–55 (2004). doi:10.1109/MCG.2004.1274061
26. Massie, T.H., Salisbury, J.K.: The phantom haptic interface: a device for probing virtual objects. In: Proceedings of the ASME Winter Annual Meeting, Symposium on Haptic Interfaces for Virtual Environment and Teleoperator Systems, vol. 55, pp. 295–300 (1994)
27. Murty, K.G.: Linear Complementarity, Linear and Nonlinear Programming. Heldermann, Berlin (1988)
28. Möller, T.: A fast triangle-triangle intersection test. J. Graph. Tools **2**(2), 25–30 (1997)
29. Oppenheim, A.V., Schafer, R., Buck, J.: Discrete-Time Signal Processing. Prentice-Hall, New York (1999)
30. Purves, D., Augustine, G.J., Fitzpatrick, D., Katz, L.C., LaMantia, A.S., McNamara, J.O., Williams, S.M.: Neuroscience, vol. 235, p. 487. Sinauer, Sunderland (1997)
31. Robles-De-La-Torre, G.: The importance of the sense of touch in virtual and real environments. IEEE Multimed. **13**(3), 24 (2006)
32. Tan, H.Z., Srinivasan, M.A., Eberman, B., Cheng, B.: Human factors for the design of force-reflecting haptic interfaces. Dyn. Syst. Control **55**(1), 353–359 (1994)
33. Yoshikawa, T., Yokokohji, Y., Matsumoto, T., Zheng, X.Z.: Display of feel for the manipulation of dynamic virtual objects. J. Dyn. Syst. Meas. Control **117**, 554 (1995)
34. Zachmann, G.: Rapid collision detection by dynamically aligned dop-trees. In: Proc. of IEEE Virtual Reality Annual International Symposium; VRAIS'98, pp. 90–97 (1998)

Chapter 4
VR System Framework

This chapter presents the overall VR system and its functional blocks with their interaction. The VR system itself consist of visual, haptic and tactile renderers which display a scene, calculated by a physical model and modified by the user through haptic devices. The primary configuration of the system uses the GRAB device as haptic device. As it features in the modified version a tactile array at each end effector to stimulate the tip of thumb and index finger, the device is also used for tactile output.

4.1 Overview of the Architecture

Achieving a convincing perception of a virtual textile requires a good compromise to be reached between the need for accuracy in the material representation and the need for speed for obtaining visually realistic simulation frame rates compatible with real-time perception. These factors have to be considered both in the visual and the haptic field. However, the graphics rendering loop has different requirements compared to the haptic rendering loop in terms of refresh frequencies. As explained in Sect. 3.1.5, stability of the control loop depends on the update rate leading to a response frequency of 300–1000 Hz to ensure accurate interaction. The more the force output is delayed with respect to the actual position and stiffness of the simulated material the more energy may be generated by the loop.

A dedicated structure has therefore been defined for adapting the different frame rates required by the mechanical simulation and the haptic rendering computations. Hence, two separate computation threads were implemented: The first is a low-frequency thread for dealing with the complex large-scale simulation of the whole cloth surface, and an accurate particle system representation integrated with state-of-the-art numerical methods for achieving quantitative accuracy of the non-linear anisotropic behaviour of cloth in real-time.

The second is a high-frequency thread for computing the local data necessary for haptic rendering and for accurately sending haptic forces back to the mechanical simulation. Maintaining the update rate in the high-frequency thread requires a

G. Böttcher, *Haptic Interaction with Deformable Objects*,
Springer Series on Touch and Haptic Systems,
DOI 10.1007/978-0-85729-935-2_4, © Springer-Verlag London Limited 2011

Fig. 4.1 Sequential flow
inside the VR system

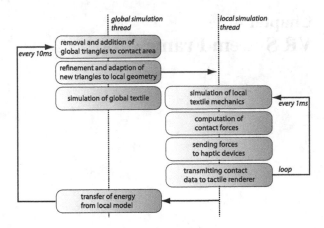

restriction to a small section of the surface for physically accurate interactions. By
taking into account the position of the cursor and its finite velocity the motion area
is well defined. Hence, a proximity region can be defined by a bounding sphere,
bounding the volume where the finger and the surface may possibly have contact.
As by Park et al. [6] proposed, restricting local consideration and computation to
be performed in the haptic loop to the parts of the surface in the bounding sphere
allows us to reduce the computation effort to a minimum. This so-called local geometry is used for the computations in this high-frequency thread. Due to the small
deformations occurring in a time frame of milliseconds, the mechanics of the local
contact is sufficiently well describing the contact of the textile. To prevent the local
geometry, i.e. the small section of the surface being in contact with the haptic cursor,
from diverging too far from the main mechanical simulation constant velocity at its
border is assumed. This implies that no forces affect the particles on the border of
the local geometry.

The flow of data within the haptic rendering is depicted in the Fig. 4.1. The architecture of the renderer is conceived to functionally separate the stages in the haptic
interaction. Apart from performance gains on multi-core systems the design allowed
to work independently on different parts relevant for the complete integration of all
hard- and software components.

4.1.1 Application Flow Inside the Framework

In the initial stage all threads are running at their dedicated update rate. The force-
feedback thread is constantly fetching new positions from the force-feedback device. These positions are processed to predict the user's motion and to estimate the
next position. At the same time the (global) textile simulation thread is computing
the deformations of the global model caused only by gravity, while the local simulation thread waits for new local geometries to simulate. At each simulation step
of the global thread it receives current and predicted fingertip positions. The global

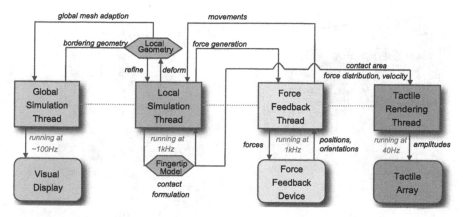

Fig. 4.2 Overview of the software architecture

thread analyses its underlying mesh for potential collisions with the fingertip for the next time step. These parts are sent to the local thread to be refined and inserted into the local simulation. Afterwards both simulation threads continue to run according to their data. With the newly added local mesh, the local thread checks if a collision has taken place in-between the last two local simulation time steps. In case of a contact the occurring deformation of the local part of the textile is computed according to the fingertip model being used. The forces at the fingertip generated during the contact are sent to the force-feedback thread. When the next global time step is reached, the changed geometry of the local mesh and the present forces in the local simulation are transferred to the global model. Figure 4.2 shows the concurrent threads within the system to create the data for each modality.

4.2 Global Simulation

The component containing the virtual textile and its physical description is the global model. It consists of a triangular mesh created in accordance to the physical dimensions and the setup of an evaluation experiment with a hanging piece of fabric (illustrated in Fig. 4.3(b)). The mesh corresponds to the particle system which is governed by the physical simulation of the textile. To each particle node within the mesh the position and the velocity is stored. As the triangles define the surface of the textile, the body masses they represent are distributed among the adjacent nodes.

As the particle system stores the physical properties of textile, its data has to be partially passed to the local simulation. Thereby, the triangles of the global model have to be identified which belong to the contact area. If triangles are below a certain distance with respect to the finger and its predicted position for the next time step, the triangles are assigned to the contact area. For a fast identification process, a collision detection is used embedding the mesh structure in a bounding volume

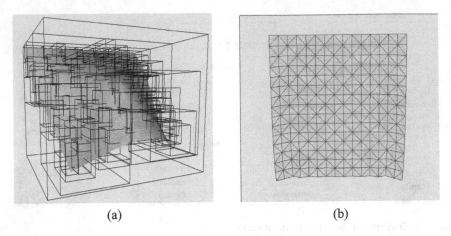

(a)   (b)

Fig. 4.3 (**a**) Dynamic AABB of global mesh. (**b**) Triangulation of the virtual textile

hierarchy of AABBs (Fig. 4.3(a)). After each simulation step the BVH is updated to regard the position changes of the particle nodes.

The physical simulation of the fabric materials is provided by a library developed at the MIRALab in a continuous effort by Volino et. al. ([8–10] etc.). The library consists of interfaces allowing to add particles with masses and to create elements building relations between the particles. These elements are used to compute the physical behaviour of continuous materials inscribed by the surrounding particles. There are different kinds of elements describing a physical portion of the material, one type computing the tensile and shearing forces as in Sect. 4.2 and another type being one- and two-dimensional expressing curvature energy representing bending stiffness. Thus altogether the elements form geometrically a line, triangle or rectangle, depending on the physical nature of the property be expressed.

4.2.1 Material Properties

For the computation of the forces created by the local elongation of the material as described in the previous section, the simulation needs to know the stresses the material will generate upon the strain. In Sect. 2.4.1 it was shown that the stresses in textiles are highly nonlinear due to the structure of the material and their obvious finite extensibility. To account for any kind of material, the stresses have to depend on the strains and its derivatives in the parameter directions.

$$\sigma_{ij}(\varepsilon_{ij}, \dot{\varepsilon}_{ij}) \quad \text{where } ij \in \{11, 12, 22\} \tag{4.1}$$

In order to simulate the behaviour of these arbitrary types of material, one has to measure the stresses σ_{ij} in various configurations of ε_{ij} and $\dot{\varepsilon}_{ij}$. Fortunately, by dealing only with fabric materials, considering every configuration in this large

Fig. 4.4 Strain-stress curve approximated by power spline

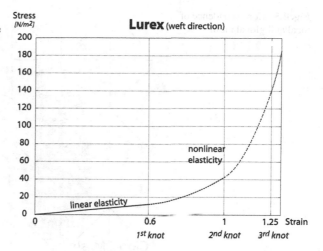

space of at least six independent directions can be omitted. As fabrics are mostly orthotropic their behaviour yields independent stress curves with respect to their strains ε_{ij} in parameter directions weft, warp, and shear. These curves are sufficient to completely describe the forces in every possible strain-stress combination.

Furthermore, the measured curves coming from the Kawabata KES-F system are approximated by a power spline as depicted in Fig. 4.4. This spline representation is used in the library for fast evaluation and its derivative of the current stresses.

$$\sigma(\varepsilon_{ij}) = A\varepsilon_{ij} + \sum_{k}\sum_{l} B_{k,l}(\varepsilon_{ij} - E_k)^l_+ \tag{4.2}$$

Whereas A is linear factor, E_k the knot vector components, and $(\varepsilon_{ij} - E_k)^l_+ = \max(0, \varepsilon_{ij} - E_k^l)$ the basis function of the spline. Basically, the resulting spline function can be seen as a series of connected polynomial functions of different degrees acting on an interval defined by the knot vector.

4.3 Local Simulation

Although many optimisations are made to the physical simulation while maintaining computational accuracy in the mechanics of the textile, it is still unable to provide the results of the update rate of the haptic device at current processor platforms. Therefore, a component is needed which fills the time gap between global model update and the force output. Moreover, this component has to be partially independent to the global model simulation process to make use of the additional processing power provided by multi-core-/multiprocessor systems.

For that reason, an intermediate model is introduced which approximates the mechanics of the highly deformable textile at the contact. It consists of a local mesh being a union of rectangular sub-meshes of the global mesh where each global triangle is refined into four triangles. As indentations of the fingers typically create a

Fig. 4.5 Correspondence of
local and global mesh

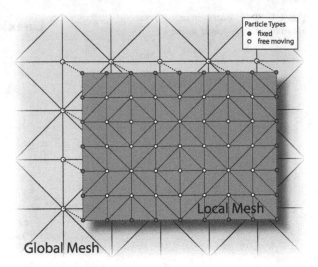

higher curvature in the textile than the global mesh is able to reproduce, the refinement at the contact allows a better approximation of the contact geometry and textile forces. Figure 4.5 illustrates the connection of local and global mesh. The physical simulation computes the deformations on the refined mesh within the small time steps and stores the energy that was induced by the user into the model until the next synchronisation. To ensure physical consistency of the deformations and the energies computed in the local step, the mechanical model used here is the same as in the global simulation. Although the local simulation is running in an independent thread, the computations must not overburden the computer in terms of haptic update rates as the underlying simulation model needs still some time for the calculations. Contrary to the global simulation being running freely without time limitations, the local simulation has to achieve real time accuracy by taking at most 1 ms for its calculations. Fulfilling these strict limitations is quite difficult as the numerical integration varies in computation time with the mesh size. In order to avoid timing issues, two run-time control algorithms have been implemented which are described in detail in Sect. 4.4.2.

4.3.1 Contact Model

In the haptic rendering of the user fingers touching the textile, the contact model is fed by the mechanics of the local simulation. Therefore, the model needs to construct the contact formulation within the small time frame as well. Unfortunately, the standard methods differ in their mechanical formulation not only yielding incompatible results with respect to the used particle model for the textile but also requiring to find the equilibrium in several iterations. The latter made them yet unusable for the constantly changing configuration of a highly deformable and very dynamic textile.

Table 4.1 Possible contact states for the finger touch

Fingers touch	No		Yes	
Friction	static / dynamic		static / dynamic	
No textile contact	I	I	II	III
One contact point	IV	V	–	–
Two contact points	VI	VII	VIII	IX

As consequence, the standard mechanical models (e.g. by Signorini) used to model the contact were not yet suited for the real-time context. More simple and more compatible methods had to be found and optimised with respect to the application. Especially, the high deformability and dynamics of the textile required special attention to be appropriately reflected inside the contact model. It became apparent that all contact states could not be solved appropriately in a single model with current computer technology. Thus contact models dedicated to each contact state have been implemented.

A prior analysis of the multi-point interaction with textiles resulted in a list of possible contact states as shown in Table 4.1. The fabric is in contact either with zero, one or two fingers. The model of the system also depends on whether the fingers touch each other or not. Note that "touch" is also used in this context if there is a fabric between the fingers, i.e. the fabric is squeezed by the two fingers. The third property determining the contact state is the kind of friction to be used, i.e. whether friction is static or dynamic.

All in all, the most important criteria in the design of the contact models are the reliability of contact detection, the stability of the force-feedback and the visual and mechanical correspondence of the action resulting from the finger's motion. These criteria are unfortunately in some aspects conflicting and a trade-off was required which considered the perception as most important.

Finally, this lead to the distinction of three cases of contact: a finger-finger, single-finger and two-finger contact case. Thus three individual algorithms were developed detecting the case and dealing with the contact states.

4.3.1.1 Finger-Finger Contact

Although fingers in this contact are not touching the textile, the illusion, however, would be disturbed if the user is not able to feel his fingers to be in contact with each other. Therefore it had not to be neglected in the force-feedback rendering.

By observing the movement of the index finger in the reference frame of the thumb, the proxy model [11] (whose concept is explained in more detail in Sect. 4.3.1.2) can be applied since only one finger is moving whereas the other is fixed. Note that it makes no difference which finger is chosen as frame of reference because the force will act on both fingers and changes its sign only. In case of an intersection of the two fingers the proxy is placed on the thumb surface in relation to the touching point of the fingers. A penalty-based method uses the penetration depth as parameter for the normal force.

To partly consider the elastic response of the fingertip, a force function was chosen to approximate the effects of deformation. Measurements of [7] yielded an exponential function describing the instantaneous elastic response of fingertip tissue. The function was used to compute the normal force F^N, but caused instabilities in the VR system due to the high stiffness. Although the stiffness of the finger is required for the contact computations, it is already included in the system by the real fingers being inside the thimbals. Therefore the compression of the textile is the only stiffness needed at this point to compute the touch force. As a compromise between stability and realism a linear spring is used with an experimentally determined stiffness of 1.25 N/mm.

The Coulomb friction model is chosen to determine the (friction) states of the haptic response when the fingers are moved against each other. One also need to address the tangential friction due to the movement of the fingers against each other. A potential next proxy position is given by the device at a new time step and the aforementioned positioning rule. Based on the distance Δx between both proxy positions, the tangential force F^T is estimated by a spring with an artificial stiffness parameter. Together with the obtained normal force and the static friction coefficient μ_s the stick-slip condition

$$F^T \leq \mu_s F^N$$

is then checked to determine the proxy movement. If the condition holds, the proxy remains on the actual position and the combined forces F^N, F^T are sent to both fingers. Otherwise the proxy is moved towards the next proxy position to fulfil the condition and the tangential force is recomputed according to the dynamic friction.

4.3.1.2 Single-Finger Contact

The algorithm is used after unsuccessful detection of the two-finger contact. This may occur when the fingers are too far apart or by having no textile between the fingers. It is run once for each finger.

A first simplification has been made by assuming the index finger in the VR system to be in contact with the back side of the fabric only whereas the thumb's contact is restricted to the front side. This assumption is justified by the mechanical setup which restricts the contact between the fingers to the palmar part of the fingertip. By the latter assumption, the reliability of the collision detection is increased as fast movement of the fingers through the thin object may rise situations were penetrations cannot be successfully detected.

If in the previous time step no has occurred, two consecutive tests are made to find possible contact points. The first one tries to find possible collision points between the finger and the mesh, this is either done for a spherical finger by considering the configuration space of the finger with respect to the textile or a collision detection for arbitrary finger mesh utilising a BVH structure (as seen in Fig. 4.6(b)).

Fig. 4.6 (**a**) Initial collision
test of finger contact.
(**b**) Collision model based on
BVH

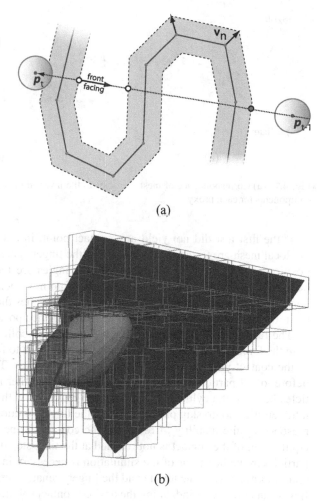

(a)

(b)

The latter represents a recent improvement to the model which strongly depends
on the processing power as the collision detection requires some of the valuable
computation time. The first approach relies on the so-called configuration space
which is defined as the set of all possible positions the finger may attain without touching the textile. In the case of a rotationally invariant shape describing
the finger, the configuration space can be constructed from the local mesh by an
offset shape with the finger radius and the vertex normal as displacement. The
new offset mesh simplifies the initial collision detection to an intersection test between the connecting line of the old reference and the new finger position. The
test disregards triangles are not front facing the last position and further omits intersection points not corresponding to the designated side of the finger. The final
contact is the closest intersection point with respect to the previous position (see
Fig. 4.6(a)).

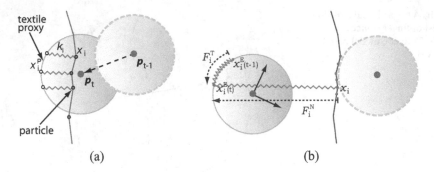

Fig. 4.7 (a) Correspondence of mesh particles to the inverted proxies. (b) Computation of force components for each proxy

If the first test did not yield any contact point, in a second try all particles of the local mesh are tested if they are inside the finger sphere surrounding the device position. As before, if the corresponding triangles are facing the wrong side, the particle will be ignored. The closest the set of point is then defined as the reference contact. In case of an empty set, the algorithm assumes that there is no contact and the reference point is identified with the actual device position.

The reference position of the finger found in the collision is now used to compute the forces occurring in the contact. First, one has to find the particles involved in the contact yielding two sets of contacting particles. The first set consists, like before, of all particles being inside the finger. The other is determined by the particles being in the cylinder connecting the reference with the device position. The latter ensures not to skip particles within the fingers motion path. All these tests are reasoned by the inability of resolving the contact at one time step. Assuming the equilibrium of the contact is not reached at the time step, tracking of the contacting particles in the iterations of the simulation is necessary in order to find a "moving" equilibrium between the textile and the finger. Penalty forces are introduced to both fingers and the textile indicating the desired contact solution. The colliding particles are related to the physically valid position on the finger, the textile proxy. The ideal position is different for the two sets of collision particles. The particles found in the device position are moved onto the finger by their vertex normal whereas the others are moved onto the device sphere by the direction of the connecting line between reference and the device position.

Springs are used to connect these particles x_i to their corresponding point x_i^E on the finger surface (see Fig. 4.7(a)). The force resultant of the force-feedback device is computed by

$$F^d = \sum_i k_i (x_i - x_i^E)$$

Each spring is associated with a contact area that is multiplied with 5 MPa/m to get the final stiffness k_i. This factor has been determined experimentally as a compromise between stability and realism. In the single finger case the textile proxies are attached only to particles of the mesh and a spring's area is calculated from the density of the material and the mass of the particle.

For friction modelling, the movement of the spring ends on the finger is controlled as follows. The spring length gives the normal force F_i^N allowing to distinguish between static and kinetic friction. And another "curved" spring, i.e. its length is measured on a surface, between the previous end and the new desired end on the finger to model the tangential force F_i^T. If the forces satisfy the stick condition equation as seen below, static friction is used for this mesh particle.

$$F_i^T < \mu_S F_i^N$$

If this condition holds, the spring end is constrained to the old position. Otherwise the equation for dynamic friction is satisfied by placing the end x_i^E at the correct position between the old and the desired position on the curved surface (see Fig. 4.7(b)).

Friction coefficients μ_s and μ_d have been measured for the movement in weft and warp, forward and reverse direction on each side of the fabric. These coefficient are interpolated at run-time for the actual movement. As the springs do not contain any information about the orientation of weft and warp directions in space, a linear least squares solver calculates them from the adjacent triangles. The direction of the fingers movement relative to the mesh particle is normalised and mapped into its local coordinate system, i.e. the difference vector the old and new spring position is projected in the weft and warp direction and used as weights for the measured friction components.

4.3.1.3 Two-Finger Case

The two-finger case considers the contact states VIII and IX and is employed if both fingers intersect. With the inability of computing the deformations occurring in the finger pinch grasp, problems arise in finding physically correct positions of the textile particles inside the intersecting finger spheres. In the single finger contact it was assumed that the finger surface does not deform under the forces produced in the textile contact. But in this case, the fingers are normally squeezed together with forces up to 10 N during the textile evaluation causing the contact surface to flatten. To approximate the behaviour, a contact plane is formed being normal to the line connecting the device positions. The contact area is then described by the circle lying on the plane of the intersecting fingers.

By computing the contact area, the occurring forces need to be applied on the portion of the textile being inside the common volume of both fingers. Since the number of textile particles within the intersecting volume is very small, their effect on the resulting contact force in relation to the contact surface they represent can become very high. Therefore, the algorithm uniformly distributes springs over contact area attached to the textile. The area the springs represent is kept constant to about

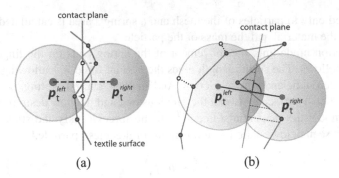

Fig. 4.8 (**a**) Textile proxies within two finger contact. (**b**) Combined contact with single and two fingers

3.35 mm^2 whereas the number of springs varies with the intersection circle. The connecting springs of previous time steps are maintained in the two finger contact as long as they are within the contact area. When the textile is sliding through the fingers, the removed springs are replaced with new springs evenly among the actual distribution. The distribution is done in two phases. In the first step a spiral function is used to iterate over the circles' area in order to find a free spot between the old springs. In a second step the remaining springs are placed in random positions in the circle by choosing a random arc length for the spiral. The ends of the springs attached in the circle are kept in plane coordinates. They therefore follow the rotation of the fingers automatically. When the circle shrinks due to the finger separation, springs outside the circle are removed.

In the last step, all mesh particles outside the contact area are treated by the single finger contact by placing textile proxies on the finger's surface using the vertex normal (Fig. 4.8). The total force created by the algorithm is distributed on the fingers by splitting the force vector into their normal and tangential component according to the contact plane. Half of the tangential force component is added to each finger's force. The direction of normal force determines the finger to which the force is added. The remaining springs outside the contact area are distributed by the previous single finger algorithm.

For the friction computation, the two-finger contact makes some simplifications resulting from the different geometry and the high tangential forces expected in this case. Instead of calculating the normal force per spring, the force is derived from the finger-finger normal force and the intersection area. The other difference is that the information about the orientation in local coordinates of the textile is provided as the spring is attached to a triangle and not to a particle. During dynamic friction the spring end slides on the contact plane and probably crosses the border of the contact area and is removed. As this would soon lead to loose the contact to the textile by the deletion all springs, deleted springs are replaced inside the circle by new ones, which preserve the contact force by having the same vector between point on the contact plane and the textile.

Fig. 4.9 Contact area of two
finger contact

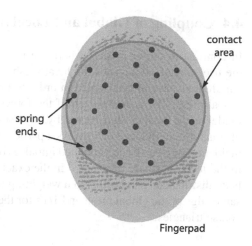

contact
area

spring
ends

Fingerpad

4.3.2 *Providing Contact Information for Tactile Rendering*

To integrate tactile feedback in the presented VR system, the force renderer needs
to compute detailed contact information. Since the tactile rendering is determined to
reproduce the signals due to surface movement, the force renderer has to keep track
of the contact in the local coordinates of the textile. Moreover, to consider the finger
orientation and the perception of edges, the renderer has to know the geometry of
the tactile array in order to distinguish between contacting and non-contacting pins.

Each actuator of the tactile array corresponds to a position on the finger stim-
ulated in case of a contact. These points have been included in the model of the
fingertip. Employing the contact area of the two finger model and the positions (in
local coordinates of the fabric) of the spring ends, the normal force at each con-
tactor pin is computed according to the spring distribution (as seen Fig. 4.9). The
estimation of the speed the contact moves along the textile is computed by a discrete
differential operator. As the contact information is sent at a lower rate of 40 Hz to
the renderer, the data resampled. The tactile renderer uses the resampled data to predict the position $\mathbf{x}_p(t)$ of the 24 actuators of each finger within the local coordinate
system of the simulated surface for the next 25 ms. The measured height profiles obtained from the KES-F system h_U and h_V creates the waveform signal of a virtual
surface

$$a_p(t) = h_U(\mathbf{U}^T \cdot \mathbf{x}_p(t)) + h_V(\mathbf{V}^T \cdot \mathbf{x}_p(t)) \quad \text{with } p \in \{1..24\} \qquad (4.3)$$

for each contacting pin. Instead, the renderer calculates the Fourier transform of
the signal and condenses the complete spectrum into the two frequencies of 40 and
320 Hz which the mechanoreceptors seem to be most sensitive to (cf. [4]). For the
motivation of the used model and detailed explanation of the tactile rendering see
[2] and [1]. But since it does not interfere with the force rendering it will not covered
beyond this section.

4.4 Coupling of Global and Local Simulation

In contrast to buffer models where the mesh or generally the data basis is fixed, the user directly influences the local mesh by his interactions. Therefore, the two simulations have to be coupled in order to propagate the local deformations to the global mesh. The accumulation of the forces, which are calculated by the contact model on the local mesh particles, by impulses w.r.t. the time step serves the purpose of transmitting induced energy. As a particle represents one third of the mass of all incident triangles, the accumulated impulses of a mesh particle which is only present in the local mesh are distributed in the exact same manner as its mass would have been distributed. This results in a weighting of 4/9 for each adjacent particle on the same edge of the global mesh and 1/18 for the opposite particles of the two incident coarse triangles.

4.4.1 Synchronisation of Simulations

To synchronise the models, different updates are done to bring the simulations into consistent state. At the beginning of the synchronisation step the global simulation needs to wait for the local thread to complete its current simulation step. Then the accumulated impulses of the local simulation are transferred as explained. As the local mesh border is constrained with the motion of the global mesh, the local particle positions are updated as well. The positions and velocity of all inner points are propagated in the other direction—from the local to the global mesh. Finally, the local geometry is updated to the new predicted situation considering the user's motion. The local geometry changes accordingly by adding (and refining) or removing (and reducing) triangles. To avoid unnecessary fluctuations in geometry size induced by misprediction or noise on the position signal, the final removal of a triangle is delayed until the triangle is not requested for several global time steps.

4.4.2 Run-Time Control

A major issue in having two concurrent simulation-threads running with different update rates is the control of the computation time. One thread has to wait for the other in order to update its internal state with the new information provided by the user interaction. As the information of the local simulation thread is directly transmitted to the user via the haptic device, the dead-time of the thread will be affecting the sensation and more importantly the stability of the force-feedback through the delay. Therefore, this thread has the highest priority and defines the time slot at which a synchronisation should occur.

By having a geometry that is changing in size at the synchronisation time, another difficulty emerges. The local geometry expands as the user moves over the

---

Algorithm 4.1 Time step control

$sim\_step \leftarrow 1\,\text{ms} \, / \, \overline{steps}$ {averages are denoted by an $\overline{overline}$}

while $deadline >$ getTime() **do**

    if $sim\_time < 1\,\text{ms}$ **then**

        simulate(sim_step)

        $sim\_time \leftarrow sim\_time + sim\_step$

    end if

    $steps\_taken \leftarrow steps\_taken + 1$

end while

$end\_time \leftarrow$ getTime()

$deadline \leftarrow deadline + 1\,\text{ms} - (end\_time - deadline) \bmod 1\,\text{ms}$

$\overline{realtime} \leftarrow \frac{7}{8} \cdot \overline{realtime} + \frac{1}{8}(end\_time - prev\_end\_time)$

$\overline{simtime} \leftarrow \frac{7}{8} \cdot \overline{simtime} + \frac{1}{8} \cdot sim\_time$

$\overline{steps} \leftarrow \frac{7}{8} \cdot \overline{steps} + \frac{1}{8} steps\_taken$

$time\_factor \leftarrow \overline{simtime} \, / \, \overline{realtime}$

$prev\_end\_time \leftarrow end\_time$

---

textile surface whereas areas which move out of the contact zone are removed. In combination with the underlying non-real-time operating system it becomes crucial to have algorithms that can adapt the computational load of the simulation to the current free CPU time in order to minimise thread dead-time and jitter in update rates.

A first algorithm was developed that was enforcing the local simulation to take a fixed computation time plus a safety margin for jitter related to the operating system. Provided by an experimentally found average triangle count ($\sim$100) the maximum number of CG iteration steps was set to ensure the numerical stability for a broad range of materials. Although the chosen maximum of 14 iterations performed well while using the system, the convergence criterion was never met as it is far too small to be reached in the real-time application. With the fixation of the computational parameters, the time required for a simulation step is determined and the only variability remains in the number of time steps to be simulated. The algorithm chooses the step size by computing a moving average of the possible number of steps estimated by the previous runs. A limit for the total time was given by the haptic device which required to get an update within a millisecond. At least one time step is computed to update the haptic device. If the computation of one step takes more time than allowed, it is evident that the CPU power is not sufficient to run the simulation in real time. But to be at least consistent within the entire physical model, the simulated time is slowed down with respect to the wall clock. Otherwise, if there is free time left, the thread waits in a busy loop for the next synchronisation.

Evidently, the algorithm described in Algorithm 4.1 is insensitive to large changes in the geometry. It needs some passes to adapt to the new requirements in simulation time. One extreme case appears in situations of very small geometries at which the *average steps* taken increase dramatically as very few computations have to be made for one time step. As a consequence, if the geometry enlarges, the algorithm takes too much steps to compute the new contact situation and the algorithm has no choice than to slow down the entire simulation. Another problem becomes visible in the opposite case of having a large contact area, which is possible in situations where the user touches the textile with both fingers at different points. As the algorithm has no option to reduce the computation time, a slowdown cannot be prevented.

Hence, an algorithm was developed by Glöckner [5] that has a better prediction of the calculation time in the next step with more parameters to influence the computational load for the numerical simulation. Finding the solution of the linear equation system consumes most of the CPU-time available. Since the employed CG solver works iteratively, it is possible to stop at a certain time, e.g. when no computation time is left. But unfortunately, the simulation library [8] had no interface to create such functionality. The only way to control the run-time is to set the maximum steps to the desired number N^{CG}. Additional control to the algorithm has been given by regarding the input variables affecting the time spent in the simulate operation. These geometry parameters are the number of particles p, the triangle face count f and the number of b^3, b^4 bend elements. Disregarding other smaller influences on run-time (e.g. CPU caches, spline order, accuracy condition) the function is of linear complexity. Thus a function t^{sim} estimating the time can be described as follows:

$$t^{sim}(p, f, b^3, b^4, N^{CG}) = c_0 + c_1 p + c_2 f + c_3 b^3 + c_4 b^4$$
$$+ N^{CG} \cdot (c_5 + c_6 p + c_7 f + c_8 b^3 + c_9 b^4)$$

where $c_0, \ldots, c_9$ are the constants to be estimated in each run on the underlying computer system.

As the inputs are known to the algorithm, it can find the constants in order to predict the time for the next simulation step t^{sim}_{n+1}. For that purpose, a least squares method is used to estimate the c_j from the geometry parameters. With the times obtained from previous simulation steps, one gets

$$\mathbf{T} := (t^{sim}_n, t^{sim}_{n-1}, \ldots)$$
$$\mathbf{g}_n := (1, p_n, f_n, b^3_n, b^4_n,$$
$$N^{CG}_n, N^{CG}_n p_n, N^{CG}_n f_n, N^{CG}_n b^3_n, N^{CG}_n b^4_n)$$
$$\mathbf{c} := (c_0, \ldots, c_9)$$

to minimise

$$\sum_i (\mathbf{g}_i^T \mathbf{c} - T_i)^2 \qquad (4.4)$$

$$\Leftrightarrow \frac{\partial}{\partial c_j} \sum_i (\mathbf{g}_i^T \mathbf{c} - T_i)^2 = 0 \quad \text{with } j = 0, \ldots, 9$$

$$\Leftrightarrow \sum_i g_{i,j}(\mathbf{g}_i^T\mathbf{c} - T_i) = 0 \quad \text{with } j = 0, \ldots, 9$$

$$\Leftrightarrow \mathbf{c}\sum_i \mathbf{g}_i^T\mathbf{g}_i = \sum_i \mathbf{g}_i\mathbf{T} \tag{4.5}$$

Finding the solution to problem (4.5) was more complex than expected because of the strong correlation between p, f, b^3 and b^4. In fact, the local geometry with f triangles has precisely $\frac{f}{2} b^4$ bend elements. That makes it impossible to simply invert the matrix given by $\mathbf{g}_i^T\mathbf{g}_i$ as it tends to be singular. Removing the dependencies would work around the problem, but to support geometries with fewer correlations another method has to be found.

In finding the solution respectively predicting the time for the next step, it does not matter which solution $\mathbf{c}$ for the underspecified linear equation system is chosen. In the literature one will find commonly used mathematical functions to solve linear algebra problems. The LAPACK software library [3] offers solution methods to such systems by finding an $\mathbf{x}$ of minimal norm that minimises $\|A\mathbf{x} - \mathbf{b}\|$ which is even well defined for A being singular. Of the available functions that fit this problem, the dgelsx function proved to be the fastest with about 56 k CPU cycles (over 260 k for the next best) by using the *AMD Core Math Library*.

In the adjustment of the $\mathbf{c}$ vector to new fitting values, one must consider how the input, namely $\mathbf{g}_i^T\mathbf{g}_i$ and $\mathbf{g}_i\mathbf{T}$, is processed. If the values for matrix and vector are solely summed up during run-time as one would prefer by looking at (4.5), the algorithm would lose its ability to react to big changes in geometry. This problem occurs when too much data is used for the estimation and is well known as overfitting. In this case both the matrix and vector components would raise to a level where floating point precision is reached and any change in the situation would have little effect on the new situation. An extreme case in which that could happen would be to leave the system running for a long time without changing the contact geometry. In order to prevent such undesired effects, the same scheme, the moving average, is employed to put more emphasis on new values rather than keeping the complete old information. In the algorithm, the scheme is used to regard values created from approximately the last 3600 time steps, earlier time steps are diminished by floating point precision.

For a perfect prediction, the algorithm would have to selectively choose values for converging to the right solution of $\mathbf{c}$ instead of minimising the norm of the problem for the current geometry. At present, phases with many changes in geometry profit from the algorithm; other phases will have to refer to the approximation of the time taken for the last CG iteration.

As seen in Algorithm 4.2, the time for one CG iteration and one simulation step with four CG iterations is predicted. This was needed to enforce numerical stability within the range of materials but could be improved by a per-material setting. Total simulation time per run is set to 1 ms for the purpose of updating the haptic device and is split into a possible amount of time steps. The remaining time is distributed on the steps by increasing the CG iterations. After one simulation step, the measurements obtained from the step are added to the aforementioned matrix and vector.

Algorithm 4.2 Improved run-time control by least squares

$\mathbf{c} \leftarrow \texttt{solveLeastSquares()}$

$t_i^{CG} = c_5 + c_6 p_i + c_7 f_i + c_8 b_i^3 + c_9 b_i^4$

$t_i^{sim} = c_0 + c_1 p_i + c_2 f_i + c_3 b_i^3 + c_4 b_i^4 + 4 t_i^{CG}$

$sim\_time \leftarrow 1$ ms

while $sim\_time > 0$ **do**

    $t^s \leftarrow \texttt{getTime()}$

    $steps \leftarrow \lfloor (deadline - t^s) / t_i^{sim} \rfloor$

    $t^{step} \leftarrow (deadline - t^s) / steps$

    $n_i \leftarrow \lfloor 4 + (t^{step} - t_i^{sim}) / t_i^{CG} \rfloor$

    if $steps < 1$ **then**

        $steps \leftarrow 1, \quad n_i \leftarrow 4$

    end if

    $sim\_step \leftarrow sim\_time / steps$

    $\texttt{simulate}(sim\_step, n_i)$

    $sim\_time \leftarrow sim\_time - sim\_step, \quad t^e \leftarrow actual\_time$

    $\texttt{addMeasurement}(p_i, f_i, b_i^3, b_i^4, n_i, t^e - t^s)$

end while

$\texttt{wait}(deadline)$

$end\_time \leftarrow \texttt{getTime()}$

$deadline \leftarrow deadline + 1$ ms $- (end\_time - deadline) \bmod 1$ ms

$\overline{realtime} \leftarrow \frac{7}{8} \cdot \overline{realtime} + \frac{1}{8}(end\_time - prev\_end\_time)$

$\overline{simtime} \leftarrow 1$ ms, $\quad time\_factor \leftarrow \overline{simtime} / \overline{realtime}$

$prev\_end\_time \leftarrow end\_time$

Although the algorithm prediction is significantly better controlled in contrast to the old one, there is still the possibility of overshooting the deadline. Therefore, a factor keeps the simulations consistent in time.

In order to achieve comparability of both algorithms, a predefined movement of user interaction has been replayed as input. As the interaction protocol also contains the timestamps, the motion is not influenced by potential simulation slowdowns. The motion describes a two-finger grasp with a following pull up of the textile and release. This action gives a good range of sizes in contact geometries. In the simulation, a coat fabric with medium bending and tensile strength was used to run the test. At each time step, the available time and the used time was logged.

Figure 4.10 shows the performance of the different algorithms with respect to the optimal utilisation of the CPU (illustrated by a dotted line). The old algorithm depicted in Fig. 4.10(a) almost always needs more time than available (about 72%) and thus slows down the overall simulation. Obviously, the fixation of the computation to 14 CG iteration steps per time step is a too hard requirement for the algorithm to wisely use the very small time slot of 1 ms. The area in the lower right corner reflects a small geometry of 8 triangles occurring for a short moment at the grasping and release of the textile. Another effect is visible by the horizontal stripes indicating that the geometry size is the main contributor to the consumed time. Because

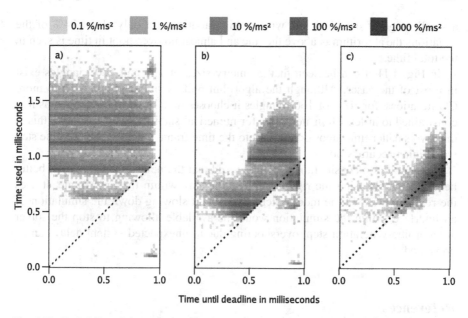

Fig. 4.10 Probability of time spent and available in the local simulation loop

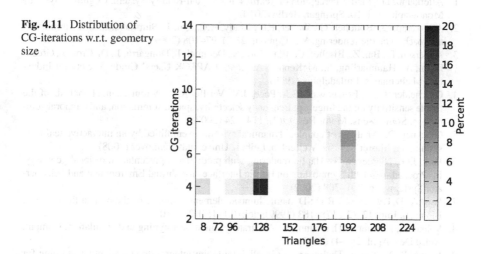

Fig. 4.11 Distribution of CG-iterations w.r.t. geometry size

of the nearly identical intensities on each if the lines, it is clear that the algorithm has no option to further reduce the computation time. Having the same numerical accuracy in both algorithms, the first algorithm was slightly modified to also allow the adjustment of CG iterations and its result is seen in Fig. 4.10(b). Here, the performance is better than before since it only takes 25% too much time (above 1 ms) and is closer to the ideal line.

With the prediction Algorithm 4.2 seen in Fig. 4.10(c) the time is still far better utilised than in the modified version of the old algorithm. It is able to stay within a small range of the ideal line without big variations. In only about 24% of the situations the algorithm is above the line and almost no overshoot in time is seen in the total time.

In Fig. 4.11 it can be seen that geometry sizes of 128 and 160 triangles exist in most of the cases. Although the algorithm prefers smaller time steps to more CG iterations, for sizes of 160 triangles it chooses mostly 10 CG iterations as it is constrained to at least four iterations for numerical stability. The reason for this is that two smaller time steps do not fit into the time frame whereas one big time step with higher accuracy does.

Taking all into consideration it can be said that the new algorithm shows better behaviour in choosing the right parameters to stay within the time frame. It uses the remaining CPU time more efficiently without slowing down the simulation. If an interface within the simulation would be available allowing to stop the solver at an arbitrary iteration step, overshooting due to unexpected system delay can be prevented.

References

1. Allerkamp, D.: Tactile Perception of Textiles in a Virtual-Reality System. Cognitive Systems Monographs, vol. 10. Springer, Berlin (2010)
2. Allerkamp, D., Böttcher, G., Wolter, F.E., Brady, A.C., Qu, J., Summers, I.R.: A vibrotactile approach to tactile rendering. Vis. Comput. 23(2), 97–108 (2007)
3. Anderson, E., Bai, Z., Bischof, C., Blackford, S., Demmel, J., Dongarra, J., Du Croz, J., Greenbaum, A., Hammarling, S., McKenney, A., et al.: LAPACK Users' Guide. Society for Industrial Mathematics, Philadelphia (1999)
4. Gescheider, G.A., Bolanowski, S.J., Pope, J.V., Verillo, R.T.: A four-channel analysis of the tactile sensitivity on the fingertip: frequency selectivity, spatial summation, and temporal summation. Somatosens. Motor Res. 19(2), 114–124 (2002)
5. Glöckner, D.: Analysis of coupled dynamical systems exemplified by an interactive real-time simulation. Master's thesis, Welfenlab, Leibniz Universität Hannover (2008)
6. Park, J.G., Niemeyer, G.: Haptic rendering with predictive representation of local geometry. In: Proceedings. 10th Symposium on Haptic Interfaces for Virtual Environment and Teleoperator Systems, pp. 201–208 (2002)
7. Pawluk, D.T.V., Howe, R.D.: Dynamic lumped element response of the human fingerpad. J. Biomech. Eng. 121(2), 178–183 (1999). doi:10.1115/1.2835100
8. Volino, P., Magnenat-Thalmann, N.: Accurate garment prototyping and simulation. Comput. Aided Des. Appl. 2(1–4) (2005)
9. Volino, P., Magnenat-Thalmann, N.: Implicit midpoint integration and adaptive damping for efficient cloth simulation. Comput. Animat. Virtual Worlds 16 (2005)
10. Volino, P., Magnenat-Thalmann, N.: Accurate anisotropic bending stiffness on particle grids. In: International Conference on Cyberworlds, CW'07, pp. 300–307 (2007)
11. Zilles, C.B., Salisbury, J.K.: A constraint-based god-object method for haptic display. In: 1995 IEEE/RSJ International Conference on Intelligent Robots and Systems 95. Human Robot Interaction and Cooperative Robots, Proceedings, vol. 3, pp. 146–151. IEEE Computer Society, Los Alamitos (1995). doi:10.1109/IROS.1995.525876

Chapter 5
Analysis of the VR System

Managing the physical simulation of textiles with haptic contact interaction for two force-feedback devices required to exploit multi-core processor architectures. But for an optimal benefit of those architectures it was required to separate the system into partly independent processes to run on separate cores. Although the simulation and the contact rendering is tightly coupled, a separation of the two functional parts was being made. By splitting the textile simulation into two simulation threads having their own geometry, it was possible to run the parts on individual processing units. These two threads responsible for their corresponding area of influence namely the global textile and the local contact, had to be synchronised as the local thread is running on a much higher rate than the other due to force-feedback rendering. Moreover, the geometry of the local simulation is only handling a small fraction of the textile thus being an incomplete physical model. The dynamic change of the local geometry imposed also timing problems difficult to handle as the computation time varies. In both threads the additional flexibility of the numerical computations utilising completely the power of the processor cores affected the timing as well.

These constraints have an influence on the dynamics of the textile. Therefore, an analysis of the developed VR system was conducted in order to assess the dynamics and the performance with respect to real time.

5.1 Effects of the Asynchronous Coupling

The effects of contact interaction with a textile or any physical object are not locally bounded. Comparable to water, deformation waves induced by the impulse from the touch emerge at the contact point and propagate viscoelastically through the entire medium.

With the separation of the local contact from the global model and the fixation, a damping is added in the aforementioned propagation. Assume that at one point in time one particle in the local mesh is excited by an impulse. Before the impulse can pass to the global mesh, first being handled in the local simulation. In the best case this adds 1 ms of delay but since the rate of the global simulation is much lower, the

G. Böttcher, *Haptic Interaction with Deformable Objects*,
Springer Series on Touch and Haptic Systems,
DOI 10.1007/978-0-85729-935-2_5, © Springer-Verlag London Limited 2011

average delay is usually higher. For a global simulation frequency of f the average delay is $0.5/f$. Advancing the global simulation consumes another $1/f$ seconds of the time before the result is passed in form of particle changes at the local border to the local simulation. In an example with $f = 100$ Hz, one gets a delay of 15 ms in the effect of the force.

Considering the particle being dependent on the linear elements formed up for the calculations of continuum mechanics, the position of the particle is determined by neighbouring particle positions and its own. The simulation method transforms this behaviour into a function which takes the positions and velocities as input to compute the new positions and velocities as output. Having now a delay within the calculations, the position of a particle is influenced not only by current configurations but also by the previous. Previous positions are fed back into the computations which in their linearised form are comparable to an IIR filter. The theory on IIR filters allows us to determine the region stability by the transformation to rational (transfer-)function. If the function has one pole not inside the unit circle, the filter is instable, i.e. when the user is moving a particle at a certain frequency, the textile will start to oscillate on its own. Avoiding these resonance frequencies in the coupling is not possible as the delays cannot be eliminated. Therefore, the only way is to reduce the effects of resonance by additionally damping the system, which moves the poles into the unit circle but decreases the dynamics of the textile.

5.1.1 Impact on Dynamics

Another analysis of the coupling will be given by a series of experiments which were conducted to measure the influence of the actions taken to achieve haptic real-time performance. The test set-up that was chosen is based on the use case of the overall system. Therefore, in the test scenario the textile was hanging down and was fixated at the top. The local geometry was also constantly fixed to a distinct rectangular area (highlighted in blue, in Fig. 5.1) and consisted of 32 global triangles. The simulated user interaction was created by applying a force at the centre of the area resembling the touch of textile. The force was perpendicular to the hanging textile and created mainly movements of the particles in the corresponding z-direction. For the analysis of dynamics the response of the textile was recorded by saving the z-coordinate of all global particles after each synchronisation with the associated simulation time.

The results were compared to a non-real-time simulation of the textile. In comparison, the real-time simulation of the textile in the VR system differs in two aspects compared. Firstly, it reduces the numerical accuracy, mainly controlled by the iterations of the CG method and the time step length to achieve the needed latency. Secondly, it incorporates a concurrent running simulation of a local region at higher mesh resolution and shorter time steps. Therefore, the reference for a comparison addressing the aforementioned characteristics is provided by three cases of a non-real-time simulation having:

(i) a higher numerical accuracy,

Fig. 5.1 Simulated touch of
the textile

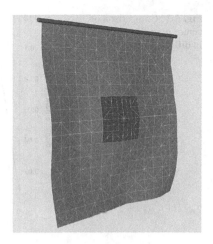

(ii) no local model (only global simulation at local resolution),
(iii) simulation combining the aspects of (i) and (ii).

Increment of numerical accuracy was accomplished on the one hand by allowing the CG solver to take up to 50 iterations for convergence. On the other hand, by reducing the time steps in the local simulation to 0.5 milliseconds but leaving synchronisation time at 1 millisecond granularity. For the first simulation variant, a constant ratio ($\frac{L_{Hz}}{G_{Hz}} = 5$) between global and local simulation rates has been set to reflect the average discrepancy in update rate of the two threads in the system. For the other reference simulations of (ii) and (iii), the local mesh was extended to the complete mesh textile which is equivalent to the global simulation at a higher mesh resolution. The difference between the latter two cases is the restriction (ii) to four CG steps as it should be comparable to the adaptive algorithm described in Sect. 4.4.2.

5.1.2 Interaction Tests

After defining the reference simulations, a variety of user interactions are possible for observation. But since the main interest lies in the perceivable effects of the systems' coupling, the experiment concentrates on two specific user inputs which have a great effect on motion of the textile and thus also on visual and haptic sense.

The first input should resemble the initial contact when the user touches the textile. More precisely, the user touches the textile surface with one finger and tries to assess the resistance against his motion perpendicular to the textile. By this motion the user is able to relate the counteracting force created by the inertia to the weight of the whole fabric with its corresponding visual dimensions. Therefore, the user can discriminate between light and heavy fabrics according to the forces produced at the touch.

Fig. 5.2 Comparison of
centre particle displacement
in different simulations

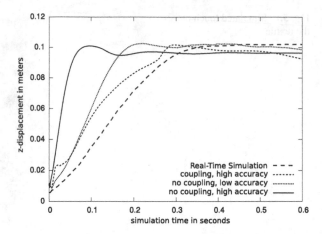

Fig. 5.3 Displacement
propagation to border in
real-time simulation

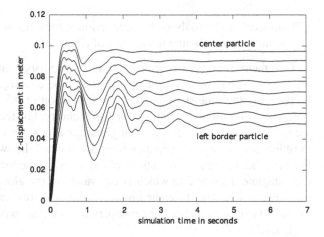

Imitating the aforementioned user action in the contact phase, a constant force of
0.3 N is applied on the center particle. The simulation stops when the new equilibrium has been reached.

The motion of the centre particle can be seen in Fig. 5.2 for the first 0.6 seconds. The simulation of the complete mesh with highest accuracy gives the best
result. It settles down at the new equilibrium within 0.15 seconds. The centre particle in the high accuracy simulation with coupling begins to move with the same
velocity but then abruptly stops at 0.01 seconds due to the fixed border. After the
synchronisation, it continues with a lower velocity to the final rest state at 0.35 seconds. It reaches the state much later than the low accuracy complete mesh variant
which arrives there in 0.3 seconds. Consequently, the real-time simulation finds its
equilibrium at the latest time of 0.4 seconds. Although for all simulations the centre particle reaches its final position quite immediately, the border particles are still
moving back and forth due to the moment given to the centre. In Fig. 5.3 the curves
for eight particles are depicted chosen from centre (topmost curve) to the left border

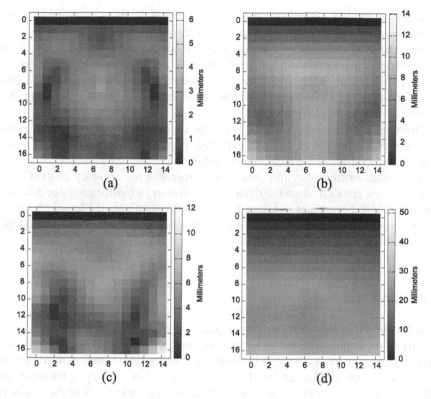

Fig. 5.4 (a) Coupled, realtime; (b) coupled, high accuracy; (c) no coupling, low accuracy; (d) no coupling, high accuracy

(lowest curve). It shows that the particles towards the border oscillate longer and their amplitude is much higher.

The investigation of the oscillation behaviour and the transmission of deformation energy was the focus for the second input. Here, the frequency response of the textile due to force inputs is to be measured. Therefore, at the centre particle a force of oscillating magnitude was applied in z-direction. It has an amplitude of 0.1 N and 4 Hz, which was chosen to allow the textile to stay in plane of the rest position (aligned within xy-plane) for all simulations. The recorded response of the particles in the z-direction during the test has been transformed in a fast Fourier Transform (FFT) with a flat-top window. A comparison in the signal strength is shown in Fig. 5.4. The grid represents the hanging fabric as seen from the front. Each grid point represents one particle with its amplitude in the main signal of 4 Hz.

The grey-scale is adjusted to meet the amplitude of the centre particle at half intensity.   The results differ significantly in amplitude at the oscillation point. For instance, in the real-time simulation an amplitude of 3 mm is reached while in the most accurate simulation it is 26 mm. This result corresponds to what has been seen before in the previous tests. The real-time simulation tends to have a high damping. In comparison to the other two simulation methods (Fig. 5.4(b) and 5.4(c)) are quite similar in results lying in-between the range of 6–7 mm. Looking more closely at the distribution of the amplitudes, one can see that there is a strong decrease in intensity outside the local area. But this artifact seems more related to the accuracy of the computations than to the coupled simulations. As the simulation is determined by the minimisation of energy which transform potential into kinetic energy, the limit to iterations of the CG solver inhibits the complete transformation resulting in a damping of the induced signal.

However, even in the high accuracy coupled simulation, the delay induced by the model is visible but not as strong as expected. It gives slightly better results as the low accuracy variant of the non-coupled simulation.

The results in frequency response have proven that the accuracy of the CG iterations has a great effect on the dynamics. Therefore, adjustments to the adaptive algorithm have been made to take advantage of the findings. By letting the algorithm favour more iterations in the CG solver over smaller simulation time steps, the dynamical behaviour is improved.

5.2 Haptic Device

Besides the VR system software, the haptic device plays an important role in the quality of the haptic perception. Without knowing the properties of the haptic device the system cannot ensure the stability and optimise the force rendering with respect to the operating range leading to vibrations and inaccuracy of forces to display. In order to prevent such unwanted effects, the characteristics of the device in terms of frequency and phase response to a force signal has to be analysed. With the knowledge a prediction of its behaviour upon force input can be used to filter frequencies which are either not displayable or which would produce instabilities.

5.2.1 Linearity Analysis

A first parameter to begin with in the analysis of the device in terms of control theoretical aspects is the relation between the input and output signals. If the input signal is amplified by the device in a non-linear manner, one has to take different approaches than the methods used for linear systems to analyse the stability. Therefore, the linearity of the device is crucial in the analysis and needs here to be measured in terms of force signals.

Inside the device the input signal is transformed within the control loop into a current for the motors. The force sensor samples the force applied on the user's finger and feeds it back into the loop. Hence, the linearity of the motors has to be considered generating torques with respect to their current and the force sensor.

The first experiment addresses the range of forces applicable to the user and the linearity of the motors. A set of defined weights were attached in different permutations to the force plate of the right arm. A test program running on the main

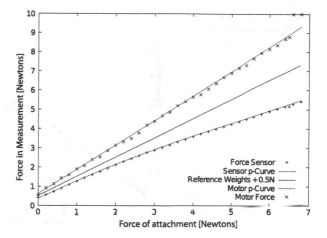

Fig. 5.5 Linearity of the device

computer had to maintain a position of 10 cm right of the centre of the device's workspace corresponding to the right end of a hanging textile within the VR system. In this position the two motors controlling the yaw of the arm are at equal load whereas the third motor in axial direction is completely unloaded. By the previously described setup the maximum force of the device within the VR system's scenario can be estimated. Varying the length of the lever by moving the device in axial direction would have led to lower respectively higher results as the torques at the motors change. With the software running the weights were attached to the force plate. The software compensated the new weights by sending higher force to the device. When the equilibrium was reached, 15000 samples were taken to measure the current force. The sampling was done at the rate of the control loop. The open velocity loop was activated to avoid the disturbances in the measurement by the input of the force sensor.

The maximum measured output force in this setup was 6.7 N including the weight of the box (0.3 N) containing the weights. To reach the maximum force, the motors had to create a torque of 0.17 N m each whereas the maximum continuous torques of the motors are specified as 0.184 N m by their technical data sheet. Moreover, the motors can attain torques of up to 2.5 N m for short time. Those peaks are limited by the motor driver used for the GRAB (cf. [1]) last up to 2.7 s and source 20 A during this period. The current corresponds to 1.2 N m on each motor yielding a maximum peak force of 47.6 N for the device provided that the difference in the torques (measured and technical maximum) accounts for the arm weight. Although the motors' limits are quite high, the device has a lower limit in the mechanical structure. The encoders in the gimbal measuring the angle of the fingers are restricted to loads of 10 N.

Figure 5.5 shows the results of the test. The black line dividing the measurements into an upper and lower section is denoting the real weights attached to the force plate. In the upper region are the force values sent to the device by the software creating the equilibrium. Since the force is generated by the open loop the samples are directly related to the motor current. Moreover, two samples on the upper right

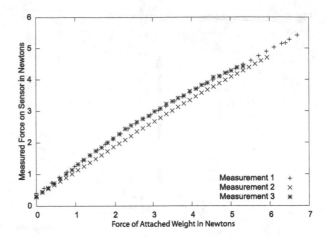

Fig. 5.6 Variation of force sensor in repeated tests

are leaving the nearly linear curve. It can be assumed that the motor driver prevents to overload the motors at continuous load. In the lower area, the samples of the force sensor are shown. This curve seems to be divided at 2.7 N into two linear sections with different slope. Repeated tests showed varying results but with a slight decrease in slope as seen in Fig. 5.6. An explanation for this could be the dependence on the orientation of the sensor, having higher accuracy in certain directions.

Finally, to judge the device in terms of linearity, one has to look for non-linearities arising from the results. These are usually expressed by the **total harmonic distortion** (THD). The THD is defined as the ratio between the energy stored in the harmonics (integer multiples of a frequency) and in the excitation frequency, i.e. it is a measure of input distortion in terms of energy. It is defined by

$$THD = \frac{\sum_{i=2}^{n} c_i^2}{c_1^2} \tag{5.1}$$

with c_1 being the amplitude of the main frequency and c_i the amplitudes of the harmonics.

A simple method to obtain the ratio is to create a sine wave signal and measure the output of the device. But in this case it was not appropriate as the sampled output signal contained too much noise related to user motion. Additional aliasing artefacts coming from the sampling have also affected the result which proved to be difficult to resolve. Therefore, another method was used by taking the values of the linearity measurement to make a Taylor approximation. With the curve describing the response to a signal given by $\hat{f}\cos(\omega t)$ a Fourier series approximation is possible to reveal the inherent harmonics and thus allows the estimation of the ratio. Firstly, least squares approximation of the data set estimates the coefficients of the Taylor polynomial and afterwards Fourier approximation gives the coefficients for the cosine basis.

$$p(x) = \sum_{i}^{n} a_i x^i \quad \overset{\text{Fourier approx.}}{\Longrightarrow} \quad p(\hat{f}\cos(\omega t)) = \sum_{i}^{m} c_i \cos(i\omega t) \tag{5.2}$$

Table 5.1 Polynomial coefficients

	Force sensor	Motor
b	$4.00293 \cdot 10^{-1}$	$4.96994 \cdot 10^{-1}$
a_1	$9.34073 \cdot 10^{-1}$	$1.25582 \cdot 10^{0}$
a_3	$-7.88355 \cdot 10^{-3}$	$1.16589 \cdot 10^{-3}$
a_5	$1.22341 \cdot 10^{-4}$	$-4.81836 \cdot 10^{-5}$
a_7	$-6.68311 \cdot 10^{-7}$	$-6.68328 \cdot 10^{-7}$

$$c_1 = \frac{35}{64}a_7\hat{f}^7 + \frac{5}{8}a_5\hat{f}^5 + \frac{3}{4}a_3\hat{f}^3 + a_1\hat{f}$$

$$c_3 = \frac{21}{64}a_7\hat{f}^7 + \frac{5}{16}a_5\hat{f}^5 + \frac{1}{4}a_3\hat{f}^3$$

$$\Rightarrow \quad c_5 = \frac{7}{64}a_7\hat{f}^7 + \frac{1}{16}\hat{f}^5$$

$$c_7 = \frac{1}{64}a_7\hat{f}^7$$

$$c_i = 0, \quad i \notin 1,3,5,7$$

It was assumed that the non-linear behaviour of the force plate is independent of the sign of force. Therefore, the curve has to be point-symmetric, which needs vanishing coefficients a_i in even i. Considering the unknown weight of force plate and the attachment of the weights, a displacement of x by b was introduced. Hence, the parameters for the approximating polynomial function were $b, a_1, a_3, a_7, \ldots$. By limiting the polynomial to a degree of 7, an average (linear) error of 0.011 N for the sensor and 0.044 N for the motors remained. A doubling of the degree has not yielded a significantly better approximation. Table 5.1 shows the results of the approximation.

With the Fourier coefficients known the computation of the total harmonic distortion as defined earlier is straight-forward. For the amplitude $\hat{f}$ the maximum force of 6.8 N was chosen covering the range of the measurement including the offset b. As the Fig. 5.5 indicated, the THD for the motors is quite low (-51.9 dB). But for the force sensor, it not so good by having only -27.7 dB between signal and harmonics. Changing the amplitudes to a lower value, the results for the sensor improve a little, -35.5 dB at 3 N and -53.6 dB for 1 N.

5.2.2 Estimation of the Sensor Resolution

In the haptic rendering, the force computation is always related to the position of the user within the VR system's environment. Depending on the algorithm used to compute the forces the position's influence varies from linear by having only springs attached to higher orders modelling volume forces. Therefore, positional resolution and estimation errors can have a strong effect on the perception in sense of smoothness of the touched object. Moreover, before the force rendering takes part, the collision detection has to determine any contacts between the interacting objects. A jittering induced by the positional noise would create highly non-linear contacts as collisions will take place and vanish instantaneously. That means the position errors can destroy history of the contact trajectory and thus the continuity when the user moves over a surface. If the haptic device suffers from noise within the positional sensing it might be beneficial to reduce the noise yielding to improved quality in haptic rendering.

Fig. 5.7 Actual measured motion from the encoder values

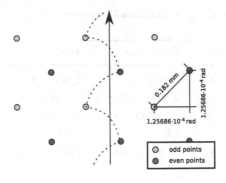

In the GRAB the position is sensed by incremental quadrature encoders. Attached to the axes, they send a square wave signal having a fixed number of impulses per revolution. In order to distinguish between clockwise and counter-clockwise, the encoders have two outputs with a phase offset of $90°$. The transition in one of the signals can be accounted for either increasing or decreasing the angle by $0.18°$. At the sensorised gimbal these values can be directly taken as the angular resolution. The other sensors are affecting the positional resolution indirectly by their attachment to the motors. Lacking any technical drawing or source code of the control software, the quantisation of the finger position has to be measured deducing the spatial resolution of the device.

But before one can estimate the resolution by considering the rotations, the centre of the logical coordinate system has to be found. By moving the arm in the axial direction to the blocking end and recording all positions while rotating, a sphere was found fitting to the positional values. With its centre it was possible to transform position data into spherical coordinates and to find the quantisation of the radius corresponding to 0.032358 mm, which coincides a 20.6 mm diameter reel at the motor controlling the axial direction.

The other two motors control respectively the sensors measure two rotational degrees of freedom of the arm as seen in Fig. 3.16 whereas the arm is fixed on a disc. Since both sensors are involved in the measurement of one rotation axis, a discrimination between the two rotations is made as follows: When the encoders notice a rotation of both motors in same direction then the disc is rotated. When both motors rotate in opposite directions then the disc is tilted. This leads to the conclusion that the difference of the encoder values is proportional to the angle between arm and ground plane and the sum of these values is proportional to the rotation at that angle. An implication of this relations between the encoder sums is that every measured point coming from an even encoder sum is surrounded by points of odd encoder sum and vice versa.

This yields for a movement in horizontal or vertical direction to a jagged curve in the sampled positions as its improbable that both encoders output changes simultaneously depicted in Fig. 5.7. By the positional recording it was found that the angles of both axis are of same proportional factor to the encoder sum or difference. A change in the values lead to changes of $1.25686 \cdot 10^{-4}$ radians. This means that the resolution with respect to the centre of the workspace is 0.182 mm. Although

Fig. 5.8 Visible structure in sensor noise

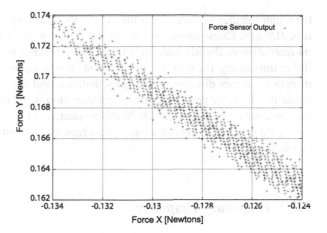

Fig. 5.9 Distances in noise

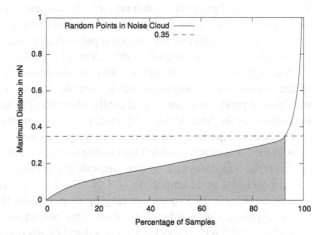

it might be sufficient for collision detection, it is not tolerable for velocity calculations deduced by the positional signal [3]. Considering the control loop sampling at 5 kHz, the velocity would be quantised to 643 mm/s without having the 500 Hz low-pass filter F_{v1} as in Fig. 3.17. But the application relies only on the positions where good resolution is needed.

Finally, to conclude the estimation of sensor resolution of the device the force sensor has to be analysed as well. Since the component values for the force vector are computed from four strain gauges, it is not simple to deduce the resolution by observing the application recording those values. Known from a previous design of the sensor having three strain gauges is that each gauge was sampled with 14 bit resolution and its force resolution was 3 mN. By looking closer at Fig. 5.8 a structure in the sensor data is visible. For the estimation of the resolution, these samples have been used to compute a convex hull. Random points were placed in this hull and their distance to the noise cloud was calculated. Accumulation of the data by sorting the distances lead to the graph of Fig. 5.9. It shows that 93% of the random points have measurement values closer than 0.35 mN. This suggests to a sensor resolution near 0.7 mN. Considering also the linearity factor a_1 of the previous experiment, its resolution is slightly reduced to 0.75 mN. Another feature worth to mention is the extraordinary form of the noise cloud. It has a pronounced expansion in one direction. In this direction the largest distance and thus its diameter is 387 mN, while the diameter perpendicular to that direction is 6.2 mN. Standard deviation in the directions are 21.8 mN and 0.36 mN respectively for orthogonal direction. Since this artifact is visible in both sensor measurements, it must be inherent to the design of the sensor.

5.2.3 Frequency Response

The key factor of the device's behaviour is the frequency response. Since the device samples positions for the VR system software and receives forces to be displayed to the user, the device transforms analog input to digital and vice versa. In this process the device is seen as a component of a control loop which transforms signals.

Although in [7] experiments and simulations conducted to analyse the response of the device, but unfortunately, since then the device by adding a heavier gimbal mechanical properties have significantly changed. Therefore, to have a complete description of the final device' response, new experiments have been conducted.

In the first experiment measuring the frequency response, a user was instructed to hold the force plate of the left arm stable with his right hand and his elbow rested on the table. The computer created sinusoidal waveforms of force output with various frequencies at an amplitude of 4 N. The output was separated into two sets containing 30 and 23 frequencies ranging from 0.5–15 Hz and 16–440 Hz. These set of frequencies were selected to observe the behaviour of the device in its operational bandwidth with respect to the closed and open control loop. Each force signal was hold for three seconds and frequencies were shuffled within a set to avoid errors in measurement due to fatigue of the user. The CPU of the main computer was hereby used as signal generator with a reference clock having a maximum sampling frequency of 2.5926 GHz.

Recorded test set contained force vectors and timestamps where only the excited directional component was used for evaluation. A sampling problem became visible justified by the design of the transmission protocol of the force sensor to the GRAB. The samples of the force sensor are not received at a fixed frequency. Moreover, the packets sent from the GRAB to the main computer featured a temporal jitter. Considering these problems, an upsampling of the data to 43.7 kHz was made by sorting the data into the sample array containing 2^{17} entries for the three seconds. The samples were placed in the entries corresponding to their time stamps. Despite the high frequency artefacts and the damping of the amplitude created by the upsampling, we can better reproduce the real signal and can easily compensate the latter artefacts.

Finally the data is processed by the flat-top windowing method as described in [4] and fast Fourier transformed. This window is used to propagate the amplitude of

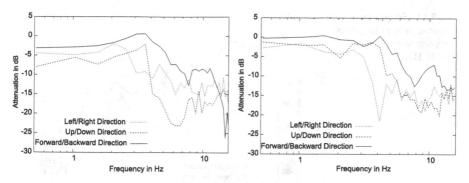

Fig. 5.10 Frequency response of the open velocity loop (*left*) and closed force loop (*right*) to 15 Hz

a free-standing frequency to the next Fourier coefficient with minimal attenuation. In using this method a maximum amplitude error of 0.11% is achieved. But contrary to the high accuracy the ability of good frequency selectivity is lost. The spectrum of a single frequency is spread by the window over about nine coefficients of the FFT whereas the highest coincides with the real amplitude.

Figure 5.10 show the response of the device in terms of signal attenuation in relation to the exciting frequency. Each curve stands for the direction of force signal and is drawn at logarithmic scale. By comparing the diagrams for the open and the closed loop a good correlation of the results between both becomes visible. One can see that they share a maximum at 4 Hz followed by a minimum at 7 Hz in response. However, the closed loop performs slightly better in the frequencies below 4 Hz as the open loop is always several dB lower. With the new results it is evident that the system has a different behaviour than before as one can see in comparing it with the plots of [1]. In the article the frequency response follows a linear segment up to 20 Hz and then falls off perfectly with 18 dB per octave to the end of the measurements at 100 Hz.

For the second set from 15 to 440 Hz, only response of the open velocity loop could be measured as seen in Fig. 5.11. In the close force loop the mechanics of the device produced loud noise and was unresponsive at these frequencies indicating the end of the stable region of force signals.

Finally, assessing the results one could say on the one hand that the signal attenuation with less than 20 dB might be sufficiently low up to 10 Hz. The user would hardly notice the small damping in the arm. Moreover, the raise in attenuation at higher frequencies has also a good side-effect. Since they are mostly related to numerical inaccuracies or measurement errors, these frequencies will be filtered out by the device improving the impression of a continuous physical behaviour. On the other hand, having an attenuation at low frequencies, which scales the force output down to the tenth of its original value, would certainly prevent the haptic rendering to be realistic in resembling the dynamics of the textile. Even at frequencies lower than 6 Hz, signals do not reach half of their amplitude. For the horizontal direction, the results show the worst response by having half amplitude already at 3.4 Hz.

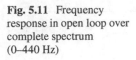

Fig. 5.11 Frequency response in open loop over complete spectrum (0–440 Hz)

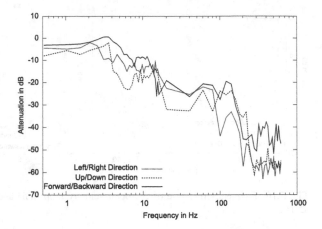

The best results are obtained in the direction towards the user. Since in this case the user's finger moves in orthogonal direction with respect to the force plate, the response is influenced by the design of the force plate.

When the finger is moving perpendicular to the plate no tangential forces occur and thus the grip is tight. In the tangential directions to the plate, the finger can start to roll or even slip out of the grip which cannot be corrected by the controller by increasing the force on the finger. Although even if the finger is fixated on the force plate by a strip, it only constrains the finger in vertical direction. In horizontal direction the finger in situations of high sudden forces regularly completely slips out of the grip. One could create a force filter that prevents such slips or increase the normal force that the device applies onto the users' finger to maintain the grip. But the latter method is not desirable as the system has already applied a high normal force of 0.5 N.

Another solution could be to change the friction coefficient between finger and force plate by applying a material with high friction. Buchholz et al. [2] conducted experiments to find materials best suited for pinch grasp. In the experiments subjects were asked to grasp a vertically oriented plate with a constant pinch force. The plate was connected to a pneumatic device which produced a ramp pulling force to provoke the slip of the plate. A strain gauge at the plate connection measured the pulling force. Different materials on the plate were used to estimate the static friction coefficient These experiments show comparable interactions with the material as in the VR system. Similar tests with the GRAB could improve the situation with grasp here.

5.2.4 Latencies

Another important property in haptic rendering is the delay between the force sent to the device and effective display at the users hand or finger. This latency between the digital input and the analog output is directly related to the phase response of the

Fig. 5.12 Eight step responses overlaid in open loop

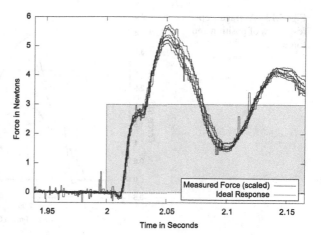

device. In control theory it is said that a linear time invariant system having solely a delay of τ possesses a linear phase and attains a 360° phase shift at a frequency of $1/\tau$. In haptics and thus for stability of the rendering, the output should correspond to the input by reducing the delay to a minimum.

To measure the delay of the device in force output, one has to know the time when the commanded force output is reached at the end-effector. Although the device features a force sensor allowing the measurement of the actual force and thus the commanded force, the acquisition of the force however is delayed as well. In fact, one has to sample the analog force again in order to obtain a measurement of the delay, which leads to another secondary delay. In any case both delays have to be considered to get an assessment of the stability.

Starting with the position sensors, the delay here is quite low as they are directly connected to an ISA bus card and sampled at the high frequency of the control loop. Accessing the registers of the optical encoders should be as low as 1 μs. Far bigger influence is expected by the force sensor. It is connected via a serial connection to the control PC. A microcontroller samples the sensor data and transfers the 100 bit packet (including start and stop bits) at 115.2 kbps and thus leading to a delay of approximately 1 ms. The network link between the control PC and the main computer of the VR system is running at 100 Mbps and accounts only for 12 μs. In the end, the main computer's OS induces delays by multi-tasking in terms of context switches and calculations.

To measure the commanded output at the force sensor, the force plate was fixated. It also enabled the operation of the GRAB without using a finger being inserted in the thimble. A step response was taken to estimate the total delay. It consisted of an alternating force of 0 to 3 N over an interval of 2 seconds perpendicular to the force plate.

Sampled forces and positions are only considered in the force component directions. In Fig. 5.12 a high correlation of the step responses in the repetitions is visible. The values measured in this experiment were scaled according to the

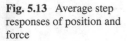

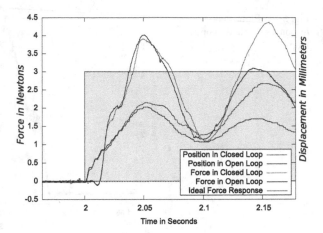

polynomial coefficients of sensor and motor obtained in the Table 5.1 to have 3 N in equilibrium. In both diagrams, the response does not immediately follow the commanded force. It is even negative about −0.15 N in the beginning. A closer look at the curves of motor positions in Fig. 5.13 indicate that this is related to the inertia and stiffness of the device mechanics as they follow the force as expected.

Drawing a conclusion of the experiment one has to consider which aspect of the latency is the most important. In defining the latency as solely the visible reaction of the device to the commanded force, the latency is depicted by the curves leaving the axis and thus leading to 1.2 ms and 7.4 ms for the motors and for the force sensor respectively.

But in this case it is more important when the commanded force is applied on the user. Therefore, the time between no force and the first moment having the correct force at the users finger is defining the latency. This value is quite different in the loops. In the open control loop the motors aim for 3 N, which is only about 2.23 N at the sensor according to the coefficients obtained in the previous experiments. Both position and force are reached after 33 ms. In the closed loop, the aim is given by the sensor and its value is reached after 41 ms, whereas the final position is attained after 49 ms.

Another definition of latency can be found in mathematics, where the derivative of the step response is the impulse response. Here, the latency is defined as time to attain the highest peak in the pulse response. Due to temporal and spatial fluctuations in the signals, it is nearly impossible to obtain an accurate derivative of the step response. But one can roughly estimate the latency by looking at the diagram. The position increases at highest in the beginning at 1.2 ms and for the forces one can see an inflection point at about 16 ms. Considering the delay of 1 ms induced by the force sensor the total latency is with respect to the used definition for the user either 6.4 ms, 32 ms, (in open loop: 40 ms) or 15 ms.

5.3 Numerical Simulation

As the simulation models mechanical behaviour of the textile and thus the reaction upon contact, its results are more or less directly perceivable through the haptic rendering. The contact model determines the state and movement of the fingers based on the textile geometry. Forces generated by the model are passed to the simulation to compute an approximated equilibrium which yields the actual configuration and forces for the haptic rendering. Regarding the connection between contact model and numerical simulation, two important properties in the simulation are apparent which strongly affect the touch perception.

The first is given by the stability of the simulation. But with the used numerical integration, the implicit midpoint rule, being an A-stable method, the simulation cannot become instable under any stiffness. However, it still requires some damping in the context of cloth simulation to achieve a stable equilibrium as shown by Volino and Magnenat-Thalmann [13]. As the materials already exhibit internal damping, the stability is not of concern here.

The second property which influences the perception is the accuracy of the simulation. While the physical model simulates the static behaviour of textiles with a decent accuracy (cf. [12]), the dynamic behaviour depends on the numerical integration. With the integration being the major expense in the computation, the implicit midpoint scheme was chosen as a compromise between accuracy and speed. Although the scheme inherently possesses some effects on accuracy by a numerically induced damping, the major contribution to errors is given by the time step size and the convergence of the CG method. Since the CG method uses most of the computation time required by the integration scheme, the number of iterations needed for convergence has effect on both the accuracy of the LES solution and the time step size. Consequently, to improve the accuracy of the simulation at the given real-time constraints a reduction in the number of iterations would have greatest effect.

The common technique in achieving better convergence and thus reducing the iterations in the CG method is preconditioning of the matrix given by the integration scheme. Therefore, different preconditioning schemes were implemented in [5], such as the Jacobi, Block–Jacobi, Richardson, and Block–Cholesky precondition schemes, in order to assess their suitability for the application in cloth simulation. For the scenario of comparison, a horizontally lying fabric fixed on two opposite sides was chosen to achieve a stable situation after few simulation steps. For the stop criterion of the preconditioned CG (PCG) method an error bound of $1 \cdot 10^{-3}$ was set which is the average error of the Taylor approximation used in the force computation. For the analysis, the number of the PCG iterations were averaged over 200 simulation steps of $1 \cdot 10^{-2}$ seconds. Additionally, the matrix was enlarged by the introduction of more particles.

Figure 5.14 shows that the Block–Cholesky scheme yields the best convergence compared to the others by limiting the average number of iterations to four. The worst convergence was given by the Block–Jacobi scheme which increases quite

Fig. 5.14 Rate of
convergence of
preconditioners

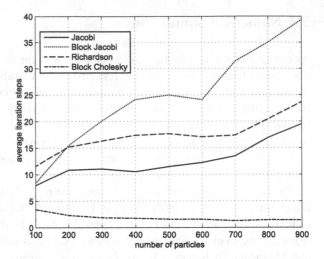

Fig. 5.15 Computational
performance

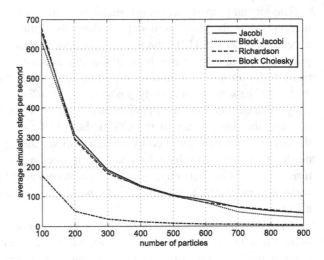

linearly to the number of particles. The remaining preconditioners perform aver-
agely by having a slight increase.

To evaluate the efficiency of the scheme in terms of computation time, the simulation steps per second were measured with respect to the problem size (Fig. 5.15).
Although the Block–Cholesky yields faster convergence, the scheme needs more
time to build up the preconditioner which limits the average number of simulation
steps per second as seen in Fig. 5.15. Jacobi and Richardson preconditions seem
to be currently the best solutions considering both rate of convergence and average
computation time. In both methods the computation time is considerably low and
both show a good rate of convergence.

Fig. 5.16   User evaluating virtual textiles with the HAPTEX VR system

5.4 Evaluation of the System

The evaluation of the VR systems rendering was separated into two steps, the objective and the subjective evaluation shown in Fig. 5.16. In the objective validation, the test procedures coming from the subjective evaluations methods were used to compare the output of the system qualitatively with the expected results. As the purpose of VR systems is to assess virtual textiles, the evaluation of the system should focus on the assessment common methods in the textile industry. But these standard methods to assess quality and suitability of fabrics are hard to be simulated due to the complex material interactions, new methods had to be defined being compatible with the VR systems manipulation capabilities. Luible et al. (see [6, 8]) defined methods considering the possible interactions of the VR system and compared their ability to assess the mechanical properties of the textile. The methods were conceived to allow only interactions of the thumb and index finger with a hanging piece of fabric fixed on a stand. Each method was explicitly devoted to the assessment of a single mechanical property. The entire evaluation procedure covered the main mechanical properties such as tensile, shear, bending, compression, friction and weight (illustrated in Fig. 5.17). From the procedures two methods, namely compression and shearing, were omitted to be evaluated since the VR system was not able to display the forces. While the compression was not modelled by the contact renderer, the shearing was covered by mechanical friction which unfortunately could not be fixed until the evaluation.

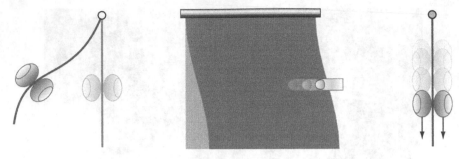

Fig. 5.17 Subjective evaluation methods (cf. [6]) for textile assessment

5.4.1 Objective Validation

Problems arise in validation of the output of the force rendering. One is founded in the complexity of the contact mechanics. Without having a reference model that is able to consider all parameters and influences involved in the contact, the output cannot be compared to the "correct" solution. The problem could be circumvented as suggested by Ruffaldi et al. [9] with an apparatus measuring the contact forces at the fingertips by a user interacting the a textile in reality. But this require special technical equipment that is not only able to measure the contact forces but also the finger position and the textile.

Another difficulty is to recreate the finger textile contact in terms of friction modelling. The friction of the skin especially at the finger varies not only from user to user but also with the material and humidity. The influencing factors have been identified by [10] yielding a high variation of the resulting friction coefficient.

Because of the inability of reproducing all the effects involved in the contact situation and thus the finding an appropriate reference solution for the output of the force render, more simple test have been conducted. These tests were defined to validate the physical correctness of some parameters separately, e.g. weight, friction and elasticity, without actual user input. The simplest test was to verify the reproduction of the textile weight. By grasping the textile with both fingertip, the total force output was validated to reflect the weight correctly.

For the verification of the correct friction modelling, an ideal fingertip movement was defined causing the contact to pass through the possible friction states. This was provided by a initial grasp of the hanging textile with a successive pull down at constant normal force. The normal force and the tangential forces were recorded until the textile has slipped through the fingertip. The ratio between normal and tangential force is shown in Fig. 5.19 to derive the friction coefficients.

Although the contact model reproduces the friction states adequately with stiff materials like Satin, it tends to overshoot in the transition from static to dynamic friction when the material is very elastic as seen in Fig. 5.18. This is caused by the spring placement of the contact model. At the point were the static friction is reached, old springs loose the contact at the top and new springs are placed at the actual fingertip at the front of the contact circle. Moreover, the textile begins to fold

Fig. 5.18 Force ratio with high stiffness

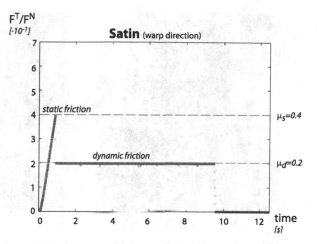

Fig. 5.19 Force ratio with low stiffness

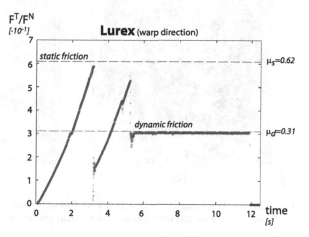

itself in front of the finger caused by the motion and its elasticity. The springs in the front therefore hold only small forces contributing to tangential force. With the removal of one back spring, the contact force is immediately reduced yielding the textile to overrelax and thus overshoot which cannot be compensated by the new springs.

5.4.2 Subjective Evaluation

Although the preceding validation of the haptic rendering indicated to be qualitatively correct in reflecting the friction and tensile properties of the textile measurements, the validation did not cover the complete system. By the creation of a virtual user, the influence of the communication with the haptic device and its response is neglected. The validation is therefore more based on an ideal haptic device being

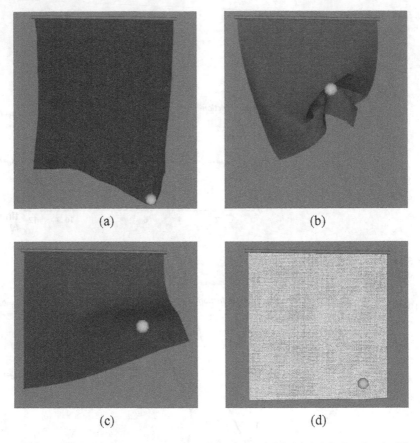

Fig. 5.20 (a) Weft knitted terry (low tensile stiffness); (b) weft knitted plain (low bend stiffness); (c) leather (high bend stiffness); (d) jute (high tensile stiffness)

perfect in response without any delay. But including the device makes it difficult to validate the obtained values as the motion of the user is not repeatable and the forces are not necessarily corresponding to the actual output of the device.

Since the purpose of the VR system was to bring the illusion of touching real fabrics to the user, it is more important to know if the system is successful in resembling the perception. Thus a subjective evaluation experiment was conducted to assess the systems performance in terms of realism. For the experiment, four test subjects were arbitrarily chosen to perform the aforementioned evaluation procedure and rate by the methods the mechanical properties. Two of them were chosen to rate the fabrics on the VR system while the other had to rate real fabrics. Due to missing modelling of compression in the VR system the compression test is omitted. Additionally, the forces occurring in the shear test are completely covered by internal friction of the haptic device in axial direction. Therefore the assessable properties were reduced to tensile stiffness, roughness, friction, bending stiffness, weight and drapability.

Table 5.2 Correlation results from the subjective evaluation (cf. [11])

Property	Tensile	Roughness	Friction	Bending
Repeatability $\varnothing$	0.98	0.93	0.67	0.18
Consistency	0.91	0.93	0.76	0.61
Realism	0.96	0.90	0.93	0.00

Five different textiles were presented twice to the subject. The subject had to rate the fabrics from 1 to 5. The upper and lower bound of each property were given by reference fabrics (see Fig. 5.20 for bending and tensile). To avoid visual cues, the evaluation was intended to be conducted blindfolded. But operating the system was hardly feasible without vision. Hence a wire-frame illustration of the textile was used to give the two subjects an equivalent environment compared to the blindfolded subjects for the real fabrics.

The analysis of the results provided the conclusion to the quality of the system in terms of haptic perception. By rating each fabric twice, a correlation between the two trials was used to estimate the "repeatability". The "consistency" in perception was obtained by correlating the ratings from the two subjects using the VR system. Finally, correlating the mean ratings of real fabrics with the virtual counterparts estimates the "realism" of the system. The summarised results are shown in Table 5.2.

With a correlation of over 0.90 for tensile stiffness and roughness, it follows that these properties were represented well by the system. The worst result is found in displaying the bending stiffness with a correlation factor of 0 in realism. The repeatability shows the similar results for a single subject. An explanation of the worse performance is probably found in the deficiency of the device displaying the very small forces occurring in the bending. The subject had therefore only the visual information by the wireframe representation to rate the bending which seemed to be not detailed enough for a reliable assessment. The system performs moderate in representing the friction of the textile by the correlation of 0.67 and 0.76 in repeatability and consistency. The high realism contradicts the previously found results. The reason is here the use of the mean values from the ratings which smoothens out the high variation among the test samples.

5.5 Conclusion

Bringing the haptic perception of computer generated objects to the user was envisioned two decades ago. Since then, researchers have worked on the different problems arising in the construction, control and simulation. The possibilities inspired discussions whether if the virtual object could ever replace the real presence. With the first commercially available device, the research has made a big leap towards this ultimate goal. Nowadays, with more matured devices capable of bring-

ing more degrees of sensation to the user and the possibility of simulating physical objects up to a certain level of accuracy in real-time, this goal got closer in
reach.

This work was aimed at tying in with the current research in haptic rendering.
With the increased computation power, the simulation of physical objects at haptic
update rates has become possible. But the power is not immediately at hand, the
new computing technology requires a rethinking in programming. Based on parallel structures, the algorithms running on the computers need to be parallel as well
to utilise the computing power. The splitting of the textile simulation into two independent processes makes an effort into the direction of exploiting the parallel
architectures. Moreover, the problems arising in the synchronisation of concurrent
simulations on both, physical and technical levels have been identified and resolved.
With the adaptive runtime control in Sect. 4.4.2, the local simulation makes most use
of the computational power available while reliably maintaining the haptic update
rate. Despite the improved accuracy in timing and resolution in the contact area, the
splitting exhibits a trade-off in the global accuracy as shown in Sect. 5.1.1 which
has also an impact on the dynamics. This problem is more visible in the simulation
of textiles which are very responsive to contact forces than soft volumetric objects.
Further observations showed that the numerical simulation is still limited by the
available computation power.

Even in the early analysis of modelling the textile contact it became very clear
that the contact models used in mechanical engineering were not suited for the
direct rendering. The approach in Sect. 4.3.1 of treating the contact cases separately allowed to create individual simplifications of the contact model with respect
to the desired situation. The inability of finding the equilibrium in one time step
posed an additional difficulty. Here, the thin structure of the textile made it necessary to predict the interaction of the two fingers, before the forces occur at the
textile.

For the force-feedback device, the main limitation lies in its maximum output
force being too low for an evaluation of virtual textiles compared to their real counterparts. Knowing the maximum now, at least allows development of a limitation in
software which would avoid accumulation of excess forces in the controller's filters
during high force periods. The non-linearities of the device and the linear factors between commanded, measured and generated forces are a matter of calibration which
can be easily fixed. The effect of the measured frequency response and the latency
on the rendering is difficult to assess as the perception is very subjective. On the
software side, there is enough potential to reduce the latency of force transmission
inside the application. By integrating the transmission control thread in the main
VR-software, the current delay of 1 ms caused by the asynchronous transmission
would be removed.

Despite all the problems and simplifications, the qualitative analysis showed that
the important mechanical properties are preserved throughout the haptic rendering.
More importantly, the subjective evaluation indicated that with the combination of
haptic and tactile rendering, some of the properties are comparable with the perception of real textiles.

References

1. Bergamasco, M., Salsedo, F., Fontana, M., Tarri, F., Avizzano, C., Frisoli, A., Ruffaldi, E., Marcheschi, S.: High performance haptic device for force rendering in textile exploration. Vis. Comput. **23**(4), 247–256 (2007)
2. Buchholz, B., Frederick, L.J., Armstrong, T.J.: An investigation of human palmar skin friction and the effects of materials, pinch force and moisture. Ergonomics **31**(3), 317–325 (1988)
3. Colgate, J.E., Brown, J.M.: Factors affecting the z-width of a haptic display. In: IEEE International Conference on Robotics and Automation, pp. 3205–3205. IEEE Press, New York (1994)
4. Gade, S., Herlufsen, H.: Use of weighting functions in DFT/FFT analysis (Part I). Brühl & Kjæær Tech. Rev. **3**, 1–22 (1987)
5. Giere, L.: Präkonditionierung des konjugierten gradienten-verfahrens in der textil-simulation. Studienarbeit, Welfenlab, Leibniz Universität Hannover (2007)
6. Luible, C., Varheenmaa, M., Magnenat-Thalmann, N., Meinander, H.: Subjective fabric evaluation. In: Proceedings of the 2007 International Conference on Cyberworlds, pp. 285–291. IEEE Computer Society, Washington (2007)
7. Marcheschi, S., Salsedo, F., Fontana, M., Tarri, F., Portillo-Rodriguez, O., Bergamasco, M., Sant'Anna, P.S.S.: High performance explicit force control for finger interaction haptic interface. In: EuroHaptics Conference, 2007 and Symposium on Haptic Interfaces for Virtual Environment and Teleoperator Systems. World Haptics 2007. Second Joint, pp. 464–469 (2007)
8. Mäkinen, M., Meinander, H., Luible, C., Magnenat-Thalmann, N.: Influence of physical parameters on fabric hand. In: Proceedings of the HAPTEX'05 Workshop on Haptic and Tactile Perception of Deformable Objects, December. University of Hannover, Hannover (2005)
9. Ruffaldi, E., Morris, D., Edmunds, T., Barbagli, F., Pai, D.K.: Standardized evaluation of haptic rendering systems. Haptic Interfaces for Virtual Environment and Teleoperator Systems, IEEE VR (2006)
10. Sivamani, R.K., Maibach, H.I.: Tribology of skin. Proc. Inst. Mech. Eng., Part J J. Eng. Tribol. **220**(8), 729–737 (2006)
11. Summers, I.: Final demonstrator and final integration report. Technical report, University of Exeter (2008)
12. Volino, P., Magnenat-Thalmann, N.: Accurate garment prototyping and simulation. Computer-Aided Design & Applications **2**(1–4) (2005)
13. Volino, P., Magnenat-Thalmann, N.: Implicit midpoint integration and adaptive damping for efficient cloth simulation. Comput. Animat. Virtual Worlds **16** (2005)

Chapter 6
Summary & Outlook

This work presented the theoretical background and the methods used in continuum mechanics and computational contact mechanics to describe the interaction of deformable bodies. With Signorini's problem a most versatile approach to contact handling has been presented. Unfortunately, the computations required to solve the contact problem with this approach is currently not possible in haptic rendering. Therefore, a simpler model was suggested in the work at hand satisfying the tight time constraints of haptics.

Furthermore a framework for the haptic rendering was developed which addresses the contact modelling between textile objects and two fingers for the first time. Although other frameworks (cf. [1, 7]) exist that allow haptic interaction between deformable objects and arbitrary tools, their application is limited to medical simulation or rigid body interaction. None of the available frameworks can be extended to textile interaction with multi-contact. Despite the common physical background, real-time simulation of fabrics is not an easy task. Due to their non-linear and highly dynamical behaviour, inaccuracies in modelling and numerical computation lead to an unrealistic behaviour. Therefore, proper simulation requires much computation power which is then unavailable for haptic rendering.

The separation of the textile simulation into a global and a local model introduced in this work, reduces the total computational load and allows the simultaneous computation of the contact forces. Although the related concept of a local buffer model to generate intermediate force updates between simulation steps of low update rates is not new (cf. [3, 4, 6]), the presented approach is superior to previous in terms of contact rendering by employing non-linear mechanics with finer mesh sizes in the local model. However, using a non-linear model in the local contact posed a problem in synchronisation, physically and technically. The physical problem has only partly been solved by the energy transfer in terms of impulses. The technical problem has been solved by an appropriate runtime control algorithm maximising computation time with respect to the time frame of 1 ms.

Moreover, as the haptic rendering depends on the contact geometry, the additional refinement of the global mesh at the contact led to improved perception of the originally smooth surface. Despite the constraints of the local model and the result-

G. Böttcher, *Haptic Interaction with Deformable Objects*,
Springer Series on Touch and Haptic Systems,
DOI 10.1007/978-0-85729-935-2_6, © Springer-Verlag London Limited 2011

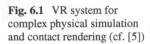

Fig. 6.1 VR system for complex physical simulation and contact rendering (cf. [5])

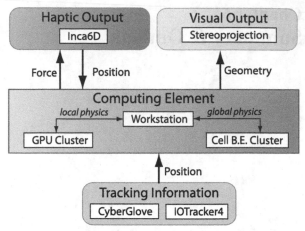

ing limitation of dynamics, the haptic perception of the mechanical characteristics of fabrics is not affected.

It became apparent in the development of the contact model that common techniques (cf. [9, 10]) were not suited for the application of touching virtual fabrics. Since these models generate forces upon the penetration of the finger, deep indentations at the contact could not be prevented without losing stability at higher contact stiffness. Consequently, the device's arms touched mechanically long before both fingers created a contact force high enough to allow grasping the virtual textile. Moreover, the models are strongly affected by the surface curvature, yielding a high variation of the valid reference position for the deformable textile and thus generating inaccurate contact forces and motion. As a result, the architecture was chosen to allow the exchange of components and to test different strategies of rendering. By the treatment of the contact in different states, new contact resolution methods were defined according to the current state. The proper tracking of the contact state not only enabled a stable contact but also the concurrent generation of tactile information of the touch used to create tactile stimuli.

While the methods implemented here are promising, especially considering the successful reconstruction of subjective evaluation procedures, some of the methods are highly specialised to the application. The real-time simulation of textiles and its haptic rendering demanded simplifications and optimisations to work convincingly with at least some of the textiles. Since specific adaptations to textiles are mainly motivated by the deficiency of computation power, parallelisation of the simulation would make more processing power available for a more general contact treatment like in Signorini's contact problem.

In [5] the development of a VR System is suggested utilising a specialised computing infrastructure for proper contact modelling. As illustrated in Fig. 6.1, the system is structured with four component types. The main component of the architecture is the computing element. This component is responsible for the processing of the data coming from the input devices. It also coordinates the information transfer to the other components. The requirement of simulating highly complex scenarios in visual as well as in haptic interactive rates necessitates the distribution of the high computational load. Therefore, the computing element is split into three components fulfilling different parts of the physical simulation. The central unit is the high performance workstation controlling the overall data flow. The two other components, a HPC-Cluster based on the Cell Processor [8] and a GPU-cluster, support the workstation in handling the high computational load for the simulation.

To be able to accelerate the numerical computations by using the two cluster components, the presented approach suggests to split the simulation into independent processes. For precise treatment of the contact, one can take the approach of Altomonte et al. [2] to solve Signorini's problem running on the GPU cluster. The global model provides the information for the visual sense and computes the dynamics of the entire physical model at visual rates. The workstation creates the link between those physical representations by energy transfers. Due to the separation it is possible to concurrently simulate the object at different rates and levels of accuracy.

References

1. Allard, J., Cotin, S., Faure, F., Bensoussan, P.J., Poyer, F., Duriez, C., Delingette, H., Grisoni, L.: Sofa-an open source framework for medical simulation. Stud. Health Technol. Inform. **125**, 13 (2006)
2. Altomonte, M., Zerbato, D., Botturi, D., Fiorini, P.: Simulation of deformable environment with haptic feedback on GPU. In: IEEE/RSJ International Conference on Intelligent Robots and Systems, 2008. IROS 2008, pp. 3959–3964 (2008)
3. Astley, O., Hayward, V.: Real-time finite elements simulation of general visco-elastic materials for haptic presentation. In: IEEE/RSJ International Conference on Intelligent Robots and Systems (IROS), IEEE Computer Society, Los Alamitos (1997)
4. Balaniuk, R.: Using fast local modelling to buffer haptic data. In: PUG99: Proceedings of the Fourth PHANTOM Users Group Workshop (1999)
5. Böttcher, G., Buchmann, R., Klein, M., Wolter, F.E.: Aufbau Eines Vr-systems zur Multimodalen Interaktion Mit Komplexen Physikalischen Modellen. In: A. Gerndt, M.E. Latoschik (eds.) Workshop der GI-Fachgruppe VR/AR, pp. 49–60. Shaker Verlag, Aachen (2009)
6. Cavusoglu, M.C., Tendick, F.: Multirate simulation for high fidelity haptic interaction with deformable. In: IEEE International Conference on Robotics and Automation, 2000. Proceedings. ICRA'00, vol. 3, pp. 2458–2464 (2000). doi:10.1109/ROBOT.2000.846397
7. Conti, F., Barbagli, F., Morris, D., Sewell, C.: Chai 3d: an open-source library for the rapid development of haptic scenes. In: Proceedings of IEEE World Haptics Conference. IEEE Computer Society, Washington (2005)
8. Kahle, J., Day, M., Hofstee, H., Johns, C., Maeurer, T., Shippy, D.: Introduction to the cell multiprocessor. IBM J. Res. Dev. **49**(4/5), 589–604 (2005)
9. Ruspini, D.C., Kolarov, K., Khatib, O.: Haptic interaction in virtual environments. In: Proceedings of the 1997 IEEE/RSJ International Conference on Intelligent Robots and Systems, 1997. IROS '97, vol. 1, pp. 128–133 (1997). doi:10.1109/IROS.1997.649024
10. Zilles, C.B., Salisbury, J.K.: A constraint-based god-object method for haptic display. In: Proceedings. 1995 IEEE/RSJ International Conference on Intelligent Robots and Systems 95. 'Human Robot Interaction and Cooperative Robots', vol. 3, pp. 146–151. IEEE Computer Society, Los Alamitos (1995). doi:10.1109/IROS.1995.525876

Index

G. Böttcher, *Haptic Interaction with Deformable Objects*,
Springer Series on Touch and Haptic Systems,
DOI 10.1007/978-0-85729-935-2, © Springer-Verlag London Limited 2011